Virus Is a Language

Virus Is a Language

AI, QAnon, COVID-19 and the New Abnormal

Chris Hables Gray

Goldsmiths Press

Copyright © 2025 Goldsmiths Press
First published in 2025 by Goldsmiths Press

Copyright © 2025 Chris Hables Gray
Goldsmiths, University of London, New Cross
London SE14 6NW

Printed and bound by CPI, UK
Distribution by the MIT Press
Cambridge, Massachusetts, and London, England

The right of the individual contributors to be identified as the authors of this work have been asserted by them in accordance with sections 77 and 78 in the Copyright, Designs and Patents Act 1988.

Every effort has been made to trace copyright holders and to obtain their permission for the use of copyright material. The publisher apologizes for any errors or omissions and would be grateful if notified of any corrections that should be incorporated in future reprints or editions of this book.

All Rights Reserved. No part of this publication may be reproduced, distributed or transmitted in any form or by any means whatsoever without prior written permission of the publisher, except in the case of brief quotations in critical articles and review and certain non- commercial uses permitted by copyright law.

A CIP record for this book is available from the British Library

ISBN 978-1-915983-37-4 Hbk
ISBN 978-1-915983-36-7 Ebk

www.gold.ac.uk/goldsmiths-press

To my sons, Corey, at the heart of my bubble throughout the COVID-19 lockdown in Santa Cruz, and Zackary, always in my thoughts, quarantined in Patagonia for the pandemic's first four months. And to Will Forest, epidemiologist, who did his duty and paid the highest price.

Contents

Acknowledgments	ix
Introducing a Spiraling World	1
Controlling Ourselves	1
Accelerating History	6
Beyond Fallacies and Illusions	8
The Relentless Calls of the Sirens	11
1 Viral Principles	17
On the Necessity of Speaking Virus	17
The Grammar of Virus	21
Viral Life Cycles	28
Language Is a Virus From Outside Space	32
2 Viral Vocabulary	39
'Ronaverse Slang	39
Technical Dictionary	43
Rhetorical Variants	49
Hosts, Vectors, Catalysts, and Controls	54
3 Viral Ideas	63
Conspiracies: Burning Books, Bodies, Minds, Forests	63
Science and COVID-19	70
Spreading Vaccines	73
Dangerous Cultural Contagions	77
4 Culturing Cultures	85
Viral Evolution and Civilization	85
Plagues in History	92
Fashions and Fads, Status and Solidarity	96
Violence Begets Violence	102

5	**Political Contagions**	107
	The Rise of Algorithmic Intelligence	107
	The Politics of Social Media	112
	QAnon and the Victory of MAGA	117
	Why Black Lives Matter, Matters	125
6	**Mutational Economies**	131
	Sacred Hunger	131
	Surveillance Capitalism	137
	Going Full Circle	141
	Spreading Innovations and Digital Proliferations	146
7	**Speaking Virus—Cyberpunk Style**	151
	The Dialect of Cyberpunk	151
	The Panoptic Present	156
	Veillance and Viruses	160
	The Affordances of Our Cyberpunk World	166
8	**Viral Futures—New Abnormals**	173
	No Normal	173
	Weird Lessons	178
	Possibilities	182
	Living in Pompeii	185
	References	195
	The Author	211
	Index	213

Acknowledgments

John Armitage, whose *Cultural Politics* issue on the COVID-19 pandemic led directly to this book. Also, Ángel Gordo, Hans Nevidal, and Karin Stürer, crucial collaborators for parts of this work.

**

Parts of this book draw from some of my articles and presentations. Specifically:

2014 "Mind control: Burning books, burning bodies, burning minds" for "Auto de Fé" art performance by Hans Nevídal, at the Deutschen Nationalbibliothek, Frankfurt, Germany.
2019 "Essay review: The threat of surveillance capitalism," *Teknokultura* 16/ 2, pp. 265–276.
2020 "Veillance society" in *The Routledge Companion to Cyberpunk*, Graham Murphy, Anna McFarlane and Lars Schmeink, eds. Routledge, pp. 362–372.
2020 "Virus is a language: The pandemic and thought control," for "Auto de Fé" art performance by Hans Nevídal, at the Frankfurt Library, Frankfurt, Germany and virtually.
2021 "Virus is a language: COVID-19 and the new abnormal," *Cultural Politics* 17/1, pp. 92–101.

Introducing a Spiraling World

Controlling Ourselves

Mitigating the risk of extinction from AI should be a global priority alongside other societal-scale risks such as pandemics and nuclear war.

—Hundreds of AI scientists and entrepreneurs (O'Brien 2023)

AI poses an existential threat and risk to health of millions.

—Hundreds of health professionals (Gregory and Hern 2023)

The newest threat to humanity in 2024 was generative artificial intelligence (AI). It seemed to come out of nowhere, first with incredible image and text manipulation, then passing bar exams and other feats. Soon, the creators and owners of the programs began warning the end of humanity might be upon us, but since then AI apps have been spreading and mutating like a virus.

Meanwhile, the biological virus SARS-CoV-2 that had dominated the world for three years had not disappeared, but the COVID-19 pandemic was over; it was now endemic, always with us, as was predicted from the start. The pandemic itself had not been unexpected after all. Nor is the next one.

And at the end of the year, QAnon achieved its greatest success ever with the election of Trump, powered in part by social media's algorithms. The massive machine learning of Algorithmic Intelligence (as Ángel Gordo and I define it) had been a long time coming, but is now here to stay. Simmering concern over several years about the racism inherent in face recognition, résumé evaluating, and other roboprocesses boiled over with the introduction of image- and text-generating programs and the dire warnings of an AI-dominated future.

At the same time, Republican politicians and right-wing billionaires began a crusade against the "woke mind virus" that was aiming at white genocide and was "worse than COVID" (Tharoor 2023). Elon Musk was particularly upset, "The woke mind virus killed my son," he declared, dead naming his trans daughter, while promising he would "destroy it" (Margaritoff 2024). The same people propagated lies about the 2024 elections (often through X/Twitter that Musk had bought) along with various QAnon theories populated by elite cannibalistic pedophile rings, lizard leaders, Jewish space lasers, and other tropes from bad science fiction, spreading them to millions through addictive AI-perfected social media algorithms.

In my 70-plus years I've learned life is a strange trip, but have to admit the last few years have been the strangest.

I went into quarantine on March 12, 2020. Sooner than most Americans but after lockdowns were implemented in Wuhan (January 23) and Lombardy (February 23). Still, it was before Santa Cruz and the Bay Area (March 15), the rest of California (March 19), or New York (March 20). On March 11, the World Health Organization (WHO) declared COVID-19 a worldwide pandemic and J and I went to San Francisco to see the show *Hamilton*. Our luck, it was the first showing cancelled, San Francisco had banned all large gatherings. On the drive back to her place in San José we realized it was time we started taking precautions as well. The next day in Santa Cruz, I told my roommate (my son C, a winemaker—not an owner or anything like that—he actually makes wine) that we would start isolating.

A few days later my friend Will Forest, an epidemiologist, sent out an email to everyone he had an address for:

You may not have heard from me in a long time. That tells you how urgent I think this message is.

You're probably accustomed to the media jumping on a news story and overcovering it and overstating the concerns. COVID-19 is the opposite of that.

I think there is no COVID control measure, large or small, that you, your family, your friends, or your local, state, or federal government will take, that you'll ever look back at and say "Oh, that was an overreaction." Quite the opposite. COVID-19 is likely to kill as many people as World War II did.

I've worked as a county health department epidemiologist for most of the last 17 years. I've been paying attention, much belatedly, to the data from China, South Korea, Italy, and Iran, as well as the bad situation in the state of Washington and the brand-new cases here in Santa Cruz, California. Hardly anyone in the media or in public office seems to understand just how bad our situation is. The numbers of cases Iran have doubled every two days for a month; most countries are on the same track, and so is Santa Cruz County. Picture a hospital where every hallway is jammed with people standing cheek to jowl like a packed elevator, waiting and hoping that they'll be the lucky person who gets the next care available. That's what actually happened in China, and they have more hospital beds per person than the U.S. does. Picture the 24/7 closure of every school, bar, restaurant, cafe, shop (except grocery stores), and so on. That's pretty much what Italy recently did, but too late; and it's what Spain and France just did. That's what we have to do, immediately, everywhere, in order to "flatten the curve"...

I'm sending you this message in the hopes that you'll contact your local and state governments to demand that they IMMEDIATELY impose the same kind of "draconian" controls that have been imposed now in Italy, France, and Spain. Next week would be TOO LATE. In this situation, the word "draconian" is kind of interchangeable with the word "inadequate." We must press for control measures that you would normally consider outrageous and beyond the pale.

You don't have anything more important than that to do today.

Since December 2019 I'd also been following COVID-19 but I didn't realize how dangerous SARS-CoV-2 was. Along with many others I'd been arguing for years that pandemics were inevitable, either xenoviruses spread by the destruction of wilderness or engineered pathogens. But one can predict something and still be surprised when it happens. Will had recently retired but he returned to work. His predictions were accurate. Some 70 to 80 million people died in World War II; as of the end of 2024 estimates collected by ourworldindata.org were of up to 40 million excess deaths from COVID-19 or effects of the pandemic such as stressed medical care and economic disruptions. Without vaccines coming incredibly fast and the lockdowns/shutdowns around the world, deaths would have been doubled or more.

As I was already obsessively reading about viruses, pandemics, and contagion in general, I leapt at the chance to write an article on "Viral Culture" for the journal *Cultural Politics*, when the call reached me on March 24. But what would I say? How could I explain what was happening?

Meanwhile, politics in the U.S. were as bad as I'd seen, part of the rise of the right around the world. QAnon was particularly mesmerizing, like watching a cobra swaying, ready to strike, its bizarre claims metastasizing around the world. Impervious to evidence, it thrived with the support of Fox News, Trump, Republicans, and the profit-maximizing algorithms of social media.

I believe in conspiracies; I've been part of some good ones. I am in several right now. I'm an active member of a revolutionary movement that goes back to the *Tao Te Ching*, the first anarchist text. But some conspiracies are more useful than others. And some conspiracies are more unhinged than I could imagine. I don't want to seem an extreme skeptic, but Queen Elizabeth *was not* an alien lizard who drank human blood. Many conspiracies are actively dangerous to society itself, and QAnon, in the context of the ever present authoritarian/populist political right, was growing exponentially more threatening until in 2024 it played a crucial role in Trump's reelection as president. It was spreading like Sars-CoV-2, not by virus-saturated micro droplets but through the tiny self-serving algorithms of the social media "meta" corporations—virtual droplets of code linking panicked minds with fearful lies.

That ideas, even cults, are viral is not a new idea. Malcolm Gladwell, the popularizer-simplifier of scientific insights proclaimed, "I'm convinced that ideas and behaviors and new products move through a population very much like a disease does. This isn't just a metaphor, in other words. I'm talking about a very literal analogy….Ideas can be contagious in exactly the same way that a virus is" (Centola 2018, p. 2). But saying this is not enough. It needs to be taken in deeply, seriously. That is the goal of this book.

Digital culture is a focus of my work. My Ph.D. was about how computers as weapons and metaphors were changing the U.S. military. Digital PSYOPs (PSYchological OPerations) and information war have been much discussed by soldiers, academics, contractors, and activists for decades, as I documented in *Postmodern War* (1991). The pure PSYOPs approach—as opposed to hybrid war where it is combined with all that messy killing—to low-intensity conflict actually doesn't work in real wars (Vietnam, Afghanistan, Iraq), but it is sometimes effective in destabilizations, regime changes, and coups. The U.S. tried out the basic principles in the overthrow of democracy in Iran (1953); they were digitalized and given a new set of justifications and technologies in Vietnam, Afghanistan, and Iraq. To use a technical CIA term, these approaches inevitably "blow back" into U.S. politics—*blow back*, the blood and tissue that blows back on you when you shoot someone close in the face. QAnon is blowback.

Russia refined massive cyber attacks against Estonia (2007), Georgia (2008), the Ukraine (2014) and other states before turning their sights on Brexit and the U.S. election of 2016. The rise of the

Russia–Trump–QAnon–social media nexus was a particularly horrifying in-my-face illustration of how virulent digital culture could be mobilized. Balanced somewhat by the same dynamics fostering radical democratic mobilizations against climate destruction (350.org, Sunrise, Extinction Rebellion, Climate Resistance) and for social justice (Idle No More, Black Lives Matter or BLM). The same reiterative machine learning processes that make large-scale, cheap, social media attacks, and recruitments possible are behind the recent explosion of generative AI programs that produce the art and texts that have shocked creatives around the world.

The science of biological viruses, reproduction, and infection rates, evolutionary changes, tipping points, and so on, resonated with what I knew already about the explosive growth of digital technology. Driven by Moore's Law and its variants, and the "creative" destruction of bleeding-edge for-profit social media technological development, it has produced surveillance capitalism (Zuboff 2019).

These phenomena don't just overlap in bodies, technologies, and cultures but in dynamics as well. Their successes and failures can be mapped on exponential curves of expansion and collapse. They spread through infection along clear vectors, catalyzed or suppressed by various factors. The similarities between biological, cultural, and digital infections were already becoming clearer when COVID-19 hit. Adam Kucharski's *The Rules of Contagion* (2020) had brought together a massive amount of evidence showing the underlying "rules" of plagues, fashions, digital disruptions, and political movements were the same. Kucharski is an epidemiologist, so his framework is biological and his focus is on outbreaks:

Whether we want an innovation to take off or an infection to decline, these are the moments we need to reach as early as possible. The moments that allow us to unravel chains of transmissions, searching for weak links, missing links, and unusual links. The moments that let us look back to work out how outbreaks really happened in the past. Then look forward to change how they happen in the future. (2020, p. 266)

To understand and deal with our intertwined political/climate/pandemic crisis, to "change how they happen in the future" we need to analyze them together. They feed each other. They take the same form and involve many of the same players. Only by drawing on viral principles—viruses are infections through information, viruses can only be understood through percentages and exponentials, viruses are sometimes like zombies from outer space—can the core dynamics of our 21st century viral polycrisis be understood, a first step towards surviving it. We need to speak Virus.

Viral emotions are integral to what is happening, as both the virus of fascism and fear-based reactions to COVID-19 make clear. But the opposite of fear, sometimes the product of fear, is bravery. Hope is something beyond that. Political, biological, and digital viruses spread not only because of their intrinsic properties but also depending on vectors, catalysts, growth mediums, and controls. Our future will be shaped by a wide range of viral forces. The key issue is contagion control, and just what mix of authoritarian control, self-control, and out-of-control (in both senses) we will end up living and dying with.

Accelerating History

Everybody knows that pestilences have a way of recurring in the world; yet somehow we find it hard to believe in ones that crash down on our heads from a blue sky.

—*Dr. Rieux in* The Plague *(Camus 1975, p. 37)*

When we follow the virus—any virus, really—we follow the fault lines of our culture.

—*Steven W. Thrasher (2022, p. 5)*

I did not enjoy sheltering in place. I am a walker and also a bit solitary. It was often an effort to go out. If it is nice, as it usually is in Santa Cruz, I try to walk a few miles daily, to buy a book, check the post office, hit the farmers market on Wednesday, or get groceries. But tell me I shouldn't go out, that my son C will shop while I "shelter in place"? Well, I did get more than a bit agoraphobic there for a few years. It still lingers.

For most of that time I only saw my son C, and after a few months my girlfriend, J. Their bubbles were slightly larger than my bubble of three but only by a few people. Twice I had socially remote drinks, once dinner, in the big back yard of K&C. My best friend R was usually there, and he and I also played disc golf up at DeLaVega a few times, going up separately and wearing masks.

After the first months of hard lockdown I shopped once a week, more in the late summer and early fall when my son was doing crush (processing the grape harvest) and working 70 hours a week. I wasn't teaching that spring but was finishing the editing of a collection of essays: *Modified: Living as a Cyborg*. So online I was in touch with my co-editors, R and my dear friend H, and with dozens of fascinating contributors from around the world, some friends but most I didn't know. Editing collections is more work for me than writing a book, so I was busy. The book had to come out, we had delayed it for over a year because of the deaths of my parents and of R's father, and the decimation of H's home Puerto Rico by hurricane Maria.

I spent hours and hours on the web working and binging on shows. I watched the first four seasons of *The Expanse* in two days. It was hard to go out. The very air seemed menacing, not to mention the idiots who still wouldn't wear a mask. There aren't many people like that in Santa Cruz, but there are some and a few dozen would meet with people from the mountain towns periodically to storm into Trader Joe's downtown, where I often shop, and make a scene. But 90% of people out and about were masked or at least wore train-robber bandanas. While I hated wearing masks, and still hate them, I wore them diligently, and even now I always carry one. I'll put it on when required or when most people are wearing them. It only seems polite.

One needs to accept that things change. This is what history is about. Most Americans are proudly ahistorical. They are more likely to fabricate the past than try and understand what really happened or even remember it. But I have always believed history is real and through our own distorted memories, and the work of imperfect historians, we can discern some of its contours. This is important because humans are not, and will never be, beyond history. History won't end until we do.

In 1989, the historian William McNeill proclaimed the "Law of Conservation of Catastrophe." He argued history was "an extraordinary, dynamic equilibrium in which triumph and disaster recur perpetually on an ever-increasing scale as our skills and knowledge grow" (McNeill 1989, p. 3). He added,

It certainly seems as though every gain in precision in the coordination of human activity and every heightening of efficiency in production were matched by a new vulnerability to breakdown. If this really is the case, then the conservation of catastrophe may indeed be a law of nature, like the conservation of energy. (pp. 11–12)

The laws in play are the ways biological, cultural, and digital viruses interact in the real world—the Language of Virus. Other relevant dynamics are dialectical, such as the human tendency to solve one problem by creating others, the mutilation/prosthesis cycle of Robbie Davis-Floyd and Joe Dumit (1998). All these processes are accelerating. Our exponentially increasing ability to manipulate the physical world has fostered many "mutilations" from greed and recklessness: massive pollution, the monetized carbon cycle, weapons proliferation, and social media that lead to dangerous individual and group behavior based on misrepresentations of bodies, both biological and political.

But society need not be driven by a quest for ever increasing profits any more than it must be organized around the life and death of this Pharaoh or Queen or some seductive religion or compelling national origin story. Yet the reigning stories do drive technology in certain directions. The roboprocesses (Besteman and Gusterson 2019) of profit and power shape the algorithmic governance, viral AI, and other digital realities that we live with now.

This is dangerous because there are now incredibly effective anti-democratic self-reinforcing systems relying on corruption-reward, coercion-punishment, and the control of information. Changing these deadly patterns requires a deeper understanding than we have used before. When one learns a new language we learn to experience the world in a new way. Virus has always been part of this world, but we usually don't notice it. We seldom see the many ways we depend on viruses for our very existence, nor the dangerous waves of biological viruses, of climate collapse, of digital innovation, of fear-driven and hate-filled political movements, until they break over our heads.

Beyond Fallacies and Illusions

A pestilence isn't a thing made to man's measure; therefore we tell ourselves that pestilence is a mere bogy of the mind, a bad dream that will pass away. But it doesn't always pass away and, from one bad dream to another, it is men who pass away.

—*Dr. Rieux in* The Plague *(Camus 1975, p. 37)*

Some people were of the opinion that living moderately and being abstemious would really help them resist the disease...Others, holding the contrary opinion, maintained that the surest medicine for such an evil disease was to drink heavily, enjoy life's pleasures, and go about singing and having fun, satisfying their appetites by any means available while laughing at everything and turning whatever happened into a joke...Of the people holding these varied opinions, not all of them died, but, by the same token not all of them survived.

—Giovanni Boccaccio (1348/2013, pp. 7–8)

My two sons, C and Z, and I spent Solstice and Christmas 2019 in Tarragona, a little Catalan city I've loved for 50 years. I could not stand being in California for the holidays because in 2018 they fell between the deaths of my atheist parents, and we didn't celebrate them at all. Ever since I moved back to California after 20 years in Oregon and Montana I'd spent them with my parents in Southern California, where my younger brother and his family also live. My mother loved the winter holidays, especially Christmas. But being half Jewish through her father she also did Hanukah and since my family and my brother's are more pagan than anything else, winter solstice. After my father died in early December, my mother carried out her desire to follow him to nothingness. Since my mother wasn't into discomfort, let alone pain, her suicide was in slow motion and didn't reach its inevitable conclusion until late January.

In Tarragona, C got strangely ill. Looking back he had all the symptoms of COVID-19, and we now know he was exposed to an early cluster while he was in Ireland before coming to Spain. His brother and I did not get sick, but we did have to cancel our trip to a cava winery–restaurant run by a friend of a friend. C was so ill that when Z and I had to move on, he stayed another week. The sweet staff at the hotel made sure he stayed hydrated and fed, checking on him often until whatever invisible beast—most likely the virus Sars-CoV-2—stopped sickening him.

So what exactly is a virus? Much of this book is about this question. It explores the basic architecture of viruses—biological, digital, and cultural—all discreet systems of information, energy, and matter exchange that depend on larger more complex systems they can commandeer for reproduction. It is also about how viruses live with us, on us, and within us. How we make viruses and how they shape us. While it is based on scholarship from history, anthropology, sociology, virology, media studies, and other disciplines and fields, this is not an academic book, just careful.

It doesn't say that SARS-CoV-2 "delights" in something, or "wants" us to open restaurants (Chen 2021a) or "wants" anything or has any "goals" at all or "switches survival strategies" or any such nonsense. Viruses do not "watch us" or "have a mind of their own" (Erill 2022). I try to use language that reflects the way evolution actually works, and it doesn't work by viruses choosing to believe and do things. Viruses don't have agency. They don't act, they react. In the same vein generative AI doesn't understand anything. It responds to stimuli, it reacts as trained and programmed. It doesn't *know*. The kind of meta-programming that would make even a crude digital consciousness possible does not exist yet.

Viruses change through evolution, driven by natural or artificial selection and programming. They mutate. Most mutations are irrelevant, many are fatal to the individual virus or ineffective algorithm, but some help them survive and reproduce, and through chance some of them actually reproduce successfully. It is a dance of *chance* and *necessity*—evolution according to Jaques Monod (1971). *Chance* produces mutations, and on rare occasions they survive and even thrive. *Necessity* is how this plays out. It either happens through the physical, biological, and cultural realities of mutation, reproduction, and proliferation...or not. The desires of viruses are irrelevant because they are nonexistent. Viruses lack the material conditions to desire.

The same cannot be said of humans. We often have more desires than is useful. But the language of Virus means we must resist our desire to use unhelpful analogies and metaphors. War is something humans do with other humans (and ants with ants, bees with bees), not with viruses or other creatures. Calling public health policies to deal with a pandemic a war only encourages attributing to the source of that pandemic, the tiny SARS-CoV-2 virus, all sorts of bizarre abilities considering it has no organs, no memory, or even nervous tissue, nothing but various spikes and other mechanisms for hijacking cells to copy itself. Viruses don't fight us. Besides, they are everywhere—the most common form of life on the planet—and we depend on them to survive and evolve.

While I've tried not to distort what viruses actually do, it has not been easy. We all slip into bad habits with our language, keeping us from seeing what's important. For example, for several decades I've been trying very hard to always be clear that humans and our works are always natural. There is no real nature-culture distinction. It is just something humans claimed to give us permission to loot the rest of nature or believe in an afterlife. Yet I still catch myself thinking of humans and our works as not natural. Of course, many cultures do not pretend humans are outside of nature, but the dominant strains of Western culture that have shaped much of my consciousness certainly have. This is a dangerous fallacy I'm trying to overcome. Saying, which is thinking, viruses have agency is equally problematic.

The Relentless Calls of the Sirens

A pandemic isn't a collection of viruses, but is a social relation among people, mediated by viruses.

—Ian Alan Paul (2020)

What makes a subject difficult to understand…is the contrast between the understanding of the subject and what most people want to see. Because of this the very things that are most obvious can become the most difficult to understand. What has to be overcome is not difficulty of the intellect but of the will.

—Ludwig Wittgenstein (1989, p. 175)

I received my first Moderna vaccination five minutes ahead of schedule, on February 5, 2021, at the Haybarn on the beautiful campus of the University of California at Santa Cruz. It went smoothly. I had a nice chat with the UCSC nurse, Brie, about detective novels, like the one clutched in my hands, and fear of shots. She gave me a very professional inoculation on my left shoulder, at the heart of the O'Flaherty tattoo my son Z and I both got in Galway some years earlier. Since Brie had an Irish background I gushed, "You really must go to Ireland, wonderful people." She gave me my vax card and I thanked her. After 15 minutes of monitoring, and one knife murder in my book, I drove home.

On my porch I made Draymond Green strongman arms and realized my shoulder was actually sore but not the injection site. Interesting. I felt good and got some work done until R invited me to a round of disc golf up at the U. Tempting. It was a spectacular day. But I realized, *I'm not immune yet*, I still shouldn't go out…probably the exercise and seeing my friend would be best for my health but I was afraid…let's say nervous. And my arm did hurt, so I didn't go. A good thing because a few minutes later I learned a module was down for my NYU course on Computers and Social Change, and I had try and get to my tech contact in New York, three hours ahead of me, hoping she hadn't left for the weekend yet, to try and keep the course on schedule.

Teaching is hard enough but mediated by digital technology it is particularly disorienting. Is it worth it? A pandemic raises all sorts of questions like this. What do we really need to survive? Food. Clothing. Housing. What do we need to thrive? Health care. Culture. Justice. At what point do we act on what we need? What makes us break from old ways when there are no guarantees the next system will be better or even stable?

This is why "tipping point" is not a perfect metaphor. There is often no tipping point into another stable system, even the "slow change" stability of the Holocene. Better to think of waves. A wave goes along driven by wind and currents, Earth tremors and passing ships, until it comes to land. Then it breaks. But that isn't the whole story. A wave can take a long time to crash and run onto the shore. In terms of Earth's health we are past the breaking point and riding the wave, riding the shockwave of 10,000 years of agriculture, 5,000 thousand years of civilization and 8 billion humans today.

There is no going back. There is no stability anymore across anyone's life, any generation, any decade. Change is so fast we can look back just a few years (2019!) and marvel at how different things are now. We ride the wave, tucked in that constantly changing corner right before it turns in on itself in roiling white foam, until we wipe out or coast gently to shore. We live in wyrd times. *Wryd*, destiny in old Germanic, implying someone or something has the power to control the future, the Fates, the Weird Sisters. *Weird.* On the surface strange, even uncanny, but when one looks deeper at the fate we are enmeshed in, clearly it is woven in part by ourselves, even if it is mainly out of our hands.

Still, we must learn, then do, what we can. One of my favorite sources of information during the Pandemic was Dr. Lucy McBride. Her weekly newsletters were thoughtfully independent. In the spring of 2022 many people were saying we were getting back to normal while others were admitting to a fear of normal (FONO), but not Dr. McBride. As she said in her March 14 *Covid-19 Update Newsletter*:

I don't have FONO right now. I have a fear of not improving on our *old* normal. Yes, I want my kids unmasked but I mostly want a better future for them and for *all* kids—a future where they can express differing opinions and engage in meaningful conversations. Where our children appreciate the diversity of people's lived experiences and still challenge the status quo. Where the world is more equitable and fair. Where public health messaging is rooted in facts and nuance and stripped of shame and fear. Where our children can protect themselves from disease and despair in tandem.

She also wrote of her earlier errors, such as too much faith in the vaccines, and how necessary it is to hold paradoxes and not collapse them. She stressed this meant thinking "about things as '*both-ands*' instead of '*either-ors*,'" adding we need to "abandon absolutism and black-or-white thinking" when dealing with our other existential crisis like "war, climate change and racism" while living through a global pandemic.

This is the thoughtful person's paradox: How much to sacrifice to the struggle for a better world? What does it mean to be maladapted to a malevolent system? Isn't the ultimate maladaptation not being able to change when change is what we need? Yet, the hardest part is figuring out what to change. Revolutions aren't easy, but destroying the old is so much easier than making a better new. Yet we must. It isn't just that a different world is possible; a better world is necessary.

We have to understand what the root of the problem is. For COVID-19, AI, QAnon, inequality, injustice, the real issue is always power. It is governance. In the face of manifest failures, who decides what changes to make or not?

Eventually, many of the mysteries of the COVID-19 pandemic will be solved. How many did this plague kill? What were its origins? Why is it so protean, with variant after variant? What policies worked? What policies failed? Some answers will never come because how much we can know has definite limits. A powerful new way of thinking about these limits was introduced in 2015 by Timothy Morton: hyperobjects. They are agglomerations that can't really be understood by us, but we can define them. Morton likes to use the example that the "styrofoam hyperobject" isn't something like all styrofoam cups at a particular time or place, but rather all styrofoam ever. Biological viruses are a hyperobject. Viral processes are an overlapping hyperobject. Virus as a language is an attempt to understand parts of the hyperobject that is the biological, digital, and cultural viral today—the parts that impact us. But Morton's concept helps us remember that we are dealing with something much more complicated than we can really imagine.

My friends JS and P live in Brooklyn at the nexus of several hospitals. When I first talked to JS soon after our lockdowns started, he told me they were having trouble sleeping because of the relentless calls of the sirens. Not mythological Greek temptresses, ambulances bringing COVID-19 patients in. In the first year of the pandemic over 30,000 New Yorkers died of the virus. I live near a fire station so I hear sirens often, but they aren't a marker of the pandemic for me. Santa Cruz never suffered the casualty rates of New York City.

Steven W. Thrasher was also in New York for the lockdown and he was also tormented by the sirens and wrote about them in his 2022 book on viruses and economics. As Jonathan Metzl explains in the Foreword, Thrasher cuts to the heart of what they mean.

> The sirens...portended not just doom but decision for people in cities like New York and in countries like the United States. Did we, the safely distanced, hear the noise and think *we* were dying? We, humans. We, citizens. We, neighbors, workers, parents, friends of friends. Or did we breathe relief and automatically think *they* are dying? They, the deserving. They, the disposable. They, the viral underclass. (p. xiii)

Thrasher's book is a brilliant report on the viral underclass. While this concept comes out of his HIV activism and scholarship, it is applicable to the whole range of viral phenomena. He shows this by following the vectors of inequality that target the most vulnerable in every society. Viruses magnify inequalities and injustices. They might not "consciously discriminate, because viruses have no consciousness" but "effects *do* discriminate against the bodies of the underclass." Some people are put "in proximity to danger" by the very ways our society is structured at the behest of the powerful. Poor people don't drown in hurricanes because water molecules have it in for them, they drown because of the "built environment of society" that keeps water away from the rich to be dumped on the poor (Thrasher 2022, pp. 5–6). In this book, I'll refer often to the viral underclass, which has to be understood in order to speak Virus, and also to Thrasher's insights about it, and how the ongoing HIV epidemic illuminates so much about the COVID-19 pandemic. He also notices,

> There is tremendous power in how, for the first time in human history, all humans on the planet have been going through some version of the same thing, at nearly the same time, with the ability to communicate globally about it. (p. 267)

I have never agreed with historians who believe there is never anything fundamentally new in human culture. They willfully ignore not just technological change but the exponential changes that have driven humanity into the current polycrises. We need new ways of understanding where we are now. Hyperobjects and the concept of the viral underclass are powerful vocabulary I am trying think with. Other ways of framing our moment are less useful for me. *Pandemicine* or *Pompeiicene*, for example, both miss the underlying viral dynamics of our age across the biological, digital, and cultural domains. The first term comes from Colin Carson, a Georgetown University biologist, and his colleague Greg Albery, who have calculated that climate change will drive an acceleration in xenoviruses because of species fleeing hotter temperatures and the further destruction of the wild. They predict at least 15,000 new spillover events "wherein viruses enter naive hosts." Carson and Albery argue that the massive changes of the Anthropocene are resulting in waves of pandemics, making it the Pandemicene (Yong 2022).

But wild virus pandemics aren't new and with improvements in vaccine creation they may never be as destructive as they were in the past, killing millions and destroying civilizations. Weaponized viruses are a different matter. How the world has responded to COVID-19 was determined more by existing digital dynamics and cultural predispositions, affordances if you will, than the purely biological. The virus itself, SARS-CoV-2, was produced by the spread of civilization (whether from lab leak or wildness destruction) and became a pandemic through globalization.

As for Pompeiicene, Pompeii wasn't a disaster produced by humans. Sure, a few philosophers realized there was a danger but it wasn't obvious, and the only special sin of the folks living on the volcano's flank was understandable ignorance. We have no excuse. From COVID-19 to the Sixth Great Extinction we know better, and yet continue on the path that has led us to now. Will Forest was right to be concerned. He died serving the community of Santa Cruz, a few years into the pandemic, killed by the virus he warned us of. The reality of the 2020s is that lifestyles and lives—even of the rich and famous—are not safe from relentless waves of change.

If it seems that our world is spiraling faster and faster that's because it is. Every day there are more people, less wildness, more science, less ecosystem stability, more carbon in the air and plastic in the sea and less time to make it right. The abnormal is normal. It is going to be new abnormal after new abnormal now. It is never going to be normal again. We are now in a world of accelerating, spiraling, viral polycrisis—and it really looks as if we are doomed. But that's the thing about seeing, often what is most important is not visible. When exponentials hold sway both the good and the bad can come and overwhelm us without warning. SARS-CoV-2 is a virus, but vaccines are a virus as well. Panic can go pandemic, but so can courage. Hope can be even more infectious than COVID-19 or QAnon. It has to be or we won't make it.

—Chris Hables Gray, Santa Cruz, California. February 17, 2025

1

Viral Principles

On the Necessity of Speaking Virus

Outbreaks are inevitable. But pandemics are optional.

—Larry Brilliant *(Mackenzie 2020, p. vi)*

And since a dead man has no substance unless one's actually seen him dead, a hundred million corpses broadcast through history are no more than a puff of smoke in the imagination.

—Dr. Reiux in The Plague *(Camus 1975, p. 38)*

If you bake you know about exponential growth. I have loved baking bread since the early 1970s when I lived at Columbae, an activist co-op at Stanford University. For my work obligation I often chose night bread baking: Two dozen loaves of Tassajara wheat bread, the Zen monastery's recipe (algorithm), three times a week. I loved the different stages that fostered the living yeast. First, feeding in warm water with honey. Some flour to create a sponge, rising and getting punched down at least twice. Then treat with more time. Add in most of the rest of the flour and the dough is workable. Knead and set to rise again. Finally, mold into loaves, lightly floured, to grow one last time. Then the oven. When the bread came out at two or three in the morning, a dozen or more housemates would appear from studying or partying or even sleeping, drawn by the thick, yeasty smell of fresh wheat bread. We'd slather butter on damp hot hunks of a sacrificial loaf and swallow them gingerly with much yelping and laughter.

Baking during the lockdown was not nearly so festive. My son C prefers mainlining protein to eating carbohydrates and my girlfriend J only visited once a week. So I ate most of the bread, the crackers, and the brownies myself.

Most people would say baking isn't a science, too much uncertainty, but then most people don't understand science. It usually doesn't deal in absolutes. It is always wrong in some ways about the most important things because we are always learning more about them. Mostly, science discovers probabilities, sometimes possibilities, but absolutely authoritative pronouncements on complicated not-fully-understood issues is not science, it is politics. So early advice on mask wearing (don't, save the masks for frontline workers), spread through surfaces (wipe down your groceries, I never did), and so on, was based on theories, not science fact. With more evidence different advice made more sense. But few things about the COVID-19 pandemic were settled, it was always changing because of new variants, shifting levels of immunity, changing protective measures, and the latest treatments. Science is a process for tracking changes while improving perception. In real life very few nontrivial issues are ever completely "solved."

New problems create new science which needs new understandings. But we don't need to rely on science for all understanding, we use art, we use philosophy, we use language. The point is Virus is a language we need to learn to help us understand the viral forces that are continually remaking our world. These aren't just biological viruses, in the digital realm and human culture there are crucial viral dynamics. COVID-19 is caused by a virus, fear is a virus, fake news and actual scientific findings can spread as viruses and, yes, in a certain light (looking out through the windows of where we might be isolating?) almost everything important seems…viral.

In her history of the 1918 influenza pandemic, *Pale Rider*, Laura Spinney explains how scientists driven by their failures, for example they first thought flu (the diminutive of influenza) was caused by a bacteria, realized that "A different narrative structure is needed, and a new language." She argues that the explosion of viral research and the new language science developed to understand epidemics, led the scientists to…

…furnish us with a vocabulary of flu—with concepts such as immune memory, genetic susceptibility and post-viral fatigue. Couched in this new language—not a poetic language, perhaps, but one that allowed you to make predictions, and to test them against the historical reports—disparate events began to appear connected, while other once obvious links atrophied and died. (2017, p. 293)

New language is often crucial for new understandings. Seeing Virus as a language is this realization writ large.

Is Virus a metaphor? An explanation? Or as Mike Davis (2020, p. 83) remarked, "...like a postmodern novel...no single narrative, but other disparate storylines racing one another to dictate a bloody conclusion"? Probably all of them. Virus is a language spoken across all domains where cybernetics rule, from not alive/not dead creatures such as the fearsome SARS-CoV-2, to the bodies of the bats and humans it infects. It also reigns in the techno-organic systems of high-tech medical research that fight COVID-19, and the distilled nuggets of fake news and scientific reports and everything else spread by digital media that infests our minds.

Considering what is happening to us biologically, digitally, and culturally, Virus seems particularly important to speak now. Let us start with an origin story. "Virus" is from the Latin, *slimy, poisonous*, as in *venom*. It moved into English in the 1500s and while first describing biological infections, within a few hundred years a virus could be a poisonous idea as well.

Biological viruses are, to quote Wikipedia, a "tiny infectious agent that reproduces inside the cells of living hosts." They are not animals, plants, fungi, or bacteria (Zimmer 2021, p. 8). They aren't even cells, just a protein coat, maybe with spikes, around some DNA or RNA (either works), which they inject into the cells of other organisms to hijack them into replicating itself. They are organic replicating nano-machines, complex selfish replicons. Some scientists say they are not alive, others that they are "on the line between chemistry and life" or living "a kind of borrowed life" (Villarreal 2008). Prions, protein chains that replicate and infect cows and humans producing fatal brain disease, are currently labeled biological but not alive. These judgments are complicated by the lack of any agreement about how to define life. Still, as more and more is learned about viruses, the case for them being living creatures grows. And they are legion.

Carl Zimmer called his book *A Planet of Viruses* because biological viruses are literally everywhere, they have extraordinary genetic diversity, and they play a fundamental role in producing much of Earth's oxygen and regulating the atmosphere. Besides all this, they are crucial for our health and integral to our genome (2021, p. 12). While there are some large viruses that can even get infected with viruses themselves, most are teeny-tiny. On average there are over 170 different species in every person's lungs;

there can be over 200 million individual viruses in a drop of water (p. 5). A liter of seawater has over a billion viruses and there are ten thousand trillion trillion viruses (roughly) in the oceans of the world. Or, to put it in another almost incomprehensible, even unhelpful, way, all the viruses in the oceans weigh as much as 75 million blue whales. Stretched out end to end, the viruses, not the whales, would span 42 million light-years (p. 57).

As of 2021, the International Committee on the Taxonomy of Viruses had identified over 9,000 virus species; there are more than a million still unidentified. Recently, researchers announced they had discovered complex ecosystems of viruses propagating 448 meters down in bedrock beneath Sweden (Bergström 2021). Viruses make up the single largest component of biomass in our biosphere, thanks to exponential growth, multiple hosts, easy mutation, and their focus on reproduction. But while sharing the same principle dynamics, digital viruses and viral processes in culture are different in important ways.

For example, computer viruses are clearly not alive. They work much as biological viruses do, but with digital algorithms instead of RNA or DNA programming. Unlike other malware such as Trojan horses and trapdoors, computer viruses are replicators, infection written real, nothing more, nothing less. They make copies of themselves that make copies of themselves. They can lurk quietly, not impacting the computer system at all, or they can reproduce insanely and overwhelm it, making the machine no more interactive than a brick. This digital virality can also be seen in machine-learning driven AI programs, simple reiterative algorithms that try many different options/mutations with the best "surviving." Viral, or Generative AI, is a focus in this book because it has an increasingly important and dangerous role in human culture, and therefore our lives.

Cultural viruses are the most complicated of all because the replication processes of culture are so much more involved than those of the organic nano-machines we call viruses or digital algorithms, but they still follow the same basic principles. In *The Rules of Contagion* (2020), the infectious disease scientist Adam Kucharsky promises that he will show...

...why financial crises are similar to sexual transmitted infections, why disease researchers found it so easy to predict games like the ice bucket challenge, and how ideas used to eradicate smallpox are helping to stop gun violence. (p. 5)

He does this but he doesn't explain why. *Why* is a deeper issue than a focus on "techniques we can use to slow down transmissions or—in the case of marketing—keep it going" (p. 5). *Why* involves recognizing that the viral is a way of understanding. Yes, we need what Kucharsky calls "outbreak science" (p. 6), but going further we need to think in new ways. We have to become fluent in Virus if we are to avoid being destroyed by these complex viral dynamics, often driven by our own choices. How can we navigate this crisis without speaking clearly about it?

This is *not* thinking like a virus. Viruses do not think. But all viral phenomena do share the same grammar.

The Grammar of Virus

The limits of my language means the limits of my world.

—Ludwig Wittgenstein (1922/1998, prop. 5.6)

Pestilence, like war, disrupts society, and silences the law.

—Tennessee Supreme Court, 1880 (quoted in Price 2022, p. 1)

Grammar is defined as the rules that govern the use and structure of a language, and therefore of the speaker's thoughts. Virus has its own grammar, revealed in the processes that govern the behavior of viruses and other microbes, the spread of human fads and fallacies, and the adoption, adaption, and rejection of new digital technologies such as the social media that has recently metastasized throughout human culture. The grammar of the viral includes basic rules and also metarules, the rules about rules.

Geneticists have successfully repurposed linguistic software, developed to learn the grammar of human languages, to improve algorithms to predict what mutations are possible in SARS-CoV-2 and which are impossible (Heaven 2021).

Their shared grammar allows different types of viral phenomena to translate across domains. In English or Spanish grammar there are rules that establish word order, punctuation, tense and aspect, determiners, connectors—which means what verbs, nouns, and pronouns do and so on. While not a full human language, Virus has its own complex grammar:

Viruses Are Infection Through Information: Infection Is Information
They are a distillation of normal information spread, the simplest form of infectious information. Biologically, viruses are just the information about how to crack a cell and reprogram it for their own reproduction. Digitally, viruses are reiterative information, such as the algorithms that create robo-processes and those driving viral AI to build big data models of human behavior. Culturally, viruses are self-reproducing systems of information living only in human brains, and thus behavior, spread in texts, symbols, music and face-to-face.

Viruses Evolve
Evolution is not a theory but a process. Systems that are not static change, as reproduction (never perfect) produces variations and some are more successful than others. Myths about evolution abound, like the wishful thinking of biologists that viruses tend to become less deadly over time, as killing off the host is a bad strategy. But this ignores history, which shows that many deadly plagues have beset humans and animals. Viruses don't plan ahead; they don't have strategies. If a virus infects quickly, especially when pre-symptomatic, it might kill almost every potential host and go extinct. Or it happens to jump into the next host before the original dies, or it spreads from the corpse, as Ebola does if the ritual washing of dead loved ones isn't stopped. When the virus can go between species, such as *yersina pestis* (bubonic plague) did and SARS-CoV-2 does, the killing off of one or more kinds of hosts is not fatal for its survival.

Just as pernicious is the idea that if a biological, digital, or cultural virus is flourishing it won't be replaced. But a new mutation still might supplant a very successful version because of its greater "fit" for the hosts and their environment or just through chance. There will always be change where there is evolution; there is always evolution where there is change.

Viruses Spread Through Networks
While biological viruses generally spread through single exposures, that is not always the case. Multiple exposures are sometimes necessary. Super-spreader events involve mass propagation through the air or, more rarely, surfaces. As with cultural viruses, the more exposure the more likely infection will occur. You are much more likely to catch cultural or biological or

digital viruses through your network of friends and colleagues than from strangers. All spread in the viral realms requires connection. Infection is a system effect.

Viruses Can Reinforce Each Other Across Domains
The sociologist Nicholas A. Christakis explains,

> ...deadly epidemics always spawn parallel epidemics of a psychological or existential nature—less tangible but equally virulent. Grief, anger, fear, denial, despair, and even anomie are not unexpected emotional reactions to the personal and collective loss in a serious outbreak of infectious disease. (2020, p. 141)

He concludes, "Contagions of germs, emotions, and behaviors can act independently or they can intersect" (p. 143). Of course, germs need a physical connection to spread, emotions and behaviors can propagate though culture, digitally or in person. The Black Death spread through Europe vectored person-to-person or rat-borne-flea-to-person, but the dancing manias, mass flagellations, and orgies spread much easier—by fear. Digitally, COVID-19 spread various simple tracking and other apps as well as some incredibly virulent nonsensical conspiracy theories with powerful political effects.

Humans Are the Viral Measure(ers)
Whether it is biological viruses, computer viruses, lies and truths, or emotions spreading, humans do the measuring. Even if the virus impacts domesticated creatures, such as our favorite zoonotic vectors the pigs and chickens, or wild ones like bats and pangolins.

Crucially, we are flawed knowing instruments. Peer pressure, anchoring effects, social and political biases, manipulation driven by ambition and fostered by cognitive dissonance and amplified by algorithms, played a major role in humanity's responses to COVID-19 on all levels, personal to global (Aronson and Tavris 2020).

Viruses Are About Percentages, Stocastic Interactions
It is all about probabilities, calculated in percentages. While individual viruses can't care about anything, we do. We don't live by percentages. It is our loved ones, or us, that get sick, hurt, or 100% dead.

Stocastic means only predictable in the aggregate—it is understanding that a chemical produces fatal cancers in 20% of those exposed to a certain level spilled, but who exactly will die is impossible to predict. Stocastic terrorism is a demagogue calling for righteous violence against a certain group, hoping it will lead to real violence but having no idea who might act, or die.

Viruses Manifest Exponential Nonlinear Growth and Collapse

The mathematician Keith Devin thinks, "Exponential growth is something that the evolutionary development of our brains did not prepare us for." Perhaps this is why it is so hard to think about exponentials, which goes beyond solving the lily pond or chess board-with-rice-grain riddles (Devin 2020). Early actions have profound consequences on this kind of growth (or collapse). The journalist Debora Mackenzie, notes "We've all seen now what exponential looks like. A short time, at the right time, matters" (2020, p. 38).

Exponentials drive many of the key dynamics of viruses, as will be detailed in the next chapter. While at least one entrepreneur-guru, Azeem Azhar, has declared this is the Exponential Age (2021), understanding involves more than investment advice. There needs to be a fundamental reordering of how we think. It is the exponential "progress" of new technologies that is driving climate change and clogging all living systems with plastics while empowering the dark magic of weapons of mass destruction, not just nuclear but also chemical and easy-to-make genocidal viruses. It is the rapidly rising risks of contagions—over 15,000 new viral transmission events are predicted over the next 50 years as the last of the wild is destroyed (Criado 2022). George Monbiot warns that our accelerating civilization, driven by new digital technologies controlled by oligarchs for their short-term gain and justified through myths about the sacred hunger of greed, is dooming humans on Earth to a death spiral as sharp as our exponential climb to 8 billion relentlessly consuming/polluting people has been (2018).

Viral Time

The processes that animate viruses happen incredibly fast…and slowly. On both ends of the spectrum they exist beyond human time. Their speed comes from their tininess and exponential reproduction. Their slowness is measured in generations and stretches back to the origins of life on this planet. Biological viruses, digital viruses, and cultural viruses seem to mutate and evolve extremely quickly, but that is an illusion caused by their high reproduction rates. When an AI deep learning system runs millions of recognition matches a minute, it is like the millions of times a day Omicron and its subvariants replicate, and that is like the billions of times a simple idea—Warriors win another NBA Championship!—propagates through human culture, morphing, latching on to other ideas and generating new stories.

Viruses have been integral to life on Earth from its beginnings. But only recently, a few million years ago, did the first human species birth one of the most complex and baroque aspects of nature on the planet—our culture. Not culture itself, many creatures have shared knowledge and values they pass down. But human culture is an outlier, qualitatively and quantitatively. Only a moment ago, in my lifetime, it spawned perhaps its weirdest extension—the digital. It is this exponential explosion of humans and our works that is behind human overpopulation and ecosystem collapse. It is all accelerating because technoculture is.

Unsurprisingly, humans have a hard time reconciling the complexities of viral time, especially policy makers. Policy is seldom made quickly, and just as rare is having it focus on the long term and not just the next election. Potential pandemics need to be confronted almost instantly by a system already in place to detect them, and that requires building a permanent infrastructure that takes into account the full spectrum of viral time.

All Models of Viruses Are Wrong

Because they are models. Because the spread of biological viruses and viral ideas and emotions such as fear, depends on how those models influence the human behaviors that shape their spread. There is a Heisenberg Uncertainty effect between models and all viral proliferation that impacts humans. Uncertainty is intrinsic to the exponential life cycle of viruses (Cockbill 2020). If that weren't enough, pandemic modeling also has to take into account viral mutations, human behavior, and the infinite complexity of the real living world. Modeling will improve, but it will never be perfect. It isn't just that a pandemic is a moving target, reality is.

When I went off to college my parents both gave me useful advice. My mom, a serious drinker, said, "Don't mix drinks. If you are drinking beer, stick to beer." My dad, a civil engineer, said, "Don't trust models." Engineers who build shit that can kill people if it fails rely on experience, science, and experiment, not unproven theories or other abstractions. The modeling of COVID-19 certainly validates my father's warning (Cepelewicz 2021a).

Viruses Need Interconnection to Survive

This is not always a symbiotic relationship. When human civilization destroys wild nature unleashing novel zootropic viruses on *Homo sapien* immune systems, it is a deadly dialectical process, with one exploitive relationship fostering another in an almost karmic cycle.

Human understandings, from the scientific to the pathologically illogical, mainly spread today through digitized networks, competing for human attention and succeeding by how well their information can infect the consciousnesses of individual bodies, their personal networks, and the body politic.

The Cybernetics of Viral Grammar

Virus are part of systems, biological or mechanistic or both (brain/culture, cyborgs). It is systems all the way down. Systems are rhetorical constructions with real consequences. It is cybernetics that links evolved and invented systems.

Viruses Are Cyborgs

Some viruses—those modified by humans or "made from scratch" as "new" polio was in 2002 (Zimmer 2021, p. 119)—are cyborgs technically, in that they are an "exogenously extended organizational complex functioning as an integrated homeostatic system unconsciously" (Clynes and Kline 1995, pp. 30-31). In practice, cyborg refers to all sorts of complex biological and machinic mergings and other systems combining the living and the dead, the invented and the evolved, flesh and metal.

Getting vaccinated, reprogramming your immune system, makes you a cyborg. Digital culture is cyborgian. We live in a cyborg society and most of us are cyborgs. Studying and making hybrid systems is cyborgology, speaking Cyborg, a language that overlaps with Virus but with its own scientific and cultural grammar and psychology (Gray 2001).

The cyborg view is not exactly the same as the viral perspective, but they certainly stand side by side. As Donna Haraway explains, "Whatever else it is the cyborg point of view is always about communication, infection, gender, genre, species, intercourse, information, and semiology" (1995: xiv). Virus and Cyborg share the deep grammar of cybernetics.

Viruses Do Not Have Agency
The problem of agency has bedeviled thinkers for as long as humans have kept track of such anxieties. Do humans have free will? Do dogs have consciousness? Is your cat torturing you on purpose? Do birds think? Do viruses? Lately, the problem has been looked at through the lens of information theory and thermodynamics. This reveals that agency requires memory to start with, and if a system doesn't have a way of keeping track of its options and choosing between them it certainly can't have agency (Kolchinsky and Wolpert 2018). This doesn't mean the memory used by AI to find matches for face recognition or manipulating human behavior into their criteria is actual choice. That isn't system memory, only task memory. Besides, agency requires levels of interactive complexity that AI, biological viruses, and infectious ideas do not and cannot exercise yet.

Viruses Are Ontologically Political
Plagues trump politics. Viruses have social implications that can't be wished away.

Viruses Must Spread
It is thrive or die...out. Viruses don't exist unless they spread and they don't spread unless they change. They can't mutate if they don't replicate. Their vocabulary is limited but powerful, starting with the verbs of four classes: vectors, catalysts, hosts, and anti-virals, discussed in the next chapter.

Viruses Have Life Cycles
Change is integral to the viral, as it is for everything that passes through time from nothingness to nothingness. But in between complexity rises and collapses in distinct patterns.

Viral Life Cycles

In our most inmate moment, as new human life emerged from old, viruses are essential to our survival. There is no us and them—just a gradually blending and shifting mix of DNA.

—Carl Zimmer (2021, p. 71)

An outbreak, like a story, should have a coherent plot.

—Philip Mortimer (quoted in Davis 2020, p. 74)

Fucking Omicron was a good lesson. I longed to go to the gym without a mask, which means go to the gym because I can't workout with a mask, I just can't. I ached to go dancing again at Moe's. I wanted to go outside without making sure I had a mask, and the latest *right* mask at that. I didn't want a return to some fictitious normal, but rather an end to that abnormal we call pandemic.

Then, in the spring of 2021, it seemed to be happening. I even went to the gym...for one week, two weeks, three weeks...That was it. Delta came. Delta was bad but in my part of California we flattened that fuck'in curve as well. My gym almost opened again. Then came Omicron.

On January 15, 2022, the tsunami caused by the eruption of Hunga Tonga-Hunga Ha-apai in the South Pacific hit Santa Cruz over 5,000 miles away, just as Omicron cases peaked. It drowned quite a few cars and trucks down at the Yacht Harbor next to the Crow's Nest, which is a great place to get Sunday breakfast watching the sailboats go out. But its impact was minimal. Not so Omicron, which shut down the town again and despite high levels of vaccinations and masks killed at least 50 'Cruzians and led to hundreds more Long COVID infections, with us still.

Among them my son C, who got it again. He lost taste and smell for a few months, for a winemaker and cook it is like losing a limb. Fortunately, these senses returned, but he still has trouble remembering lists of things. While he used to keep all his winemaking tasks in his head he now had to write them down. And he's lost his keys dozens of times while in the other ten years we lived together he might have misplaced them twice. He isn't sure which variant caused the long-term changes, out of the three waves he figures he caught. Or all three. The more times you get COVID-19 the higher the chance of Long COVID.

SARS-CoV-2 variants become important when they out compete the others and form large waves. As anyone who has spent significant time in the Pacific Ocean knows, no two waves are the same. They vary in more ways than snowflakes. Some waves can knock you on your ass. Some waves can take you to heaven for that one ride. Some waves can kill you. Waves are made up of micro-units as all viral phenomena are—molecules of air, molecules of water, little hunks of code, simple ideas, individual viruses.

We can sort of see parts of an individual virus, thanks to science, but we clearly see billions of viruses reflected in the rise and fall of infection rates, high counts in tested sewage, and excess human deaths. Elegantly visualized in abstractions of their progress, waves are always about curves.

Adrian Gore (2022) notes that popular as the Bell Curve (Gaussian distributions) is, most important dynamics follow some sort of power distribution. Like the so-called Pareto curves where 20% of something causes 80% of something, as in 20% of the people do 80% of the work. This should not be surprising. Reality, especially exponentially driven dynamics, is never regular. This is true of all viruses, not just the biological. Gore claims, without much evidence, that people prefer the Bell Curve because we are neurologically averse to inequality (although it is explicitly unequal), in favor of fairness, and they are more symmetrical. But even if true we should resist this prejudice. Reality isn't fair.

Besides, Adrian Gore's examples of Pareto curves range from 20% of Italians own 80% of the land in Italy to 25% of tweeters produced 97% of tweets to 30% of drivers cause 50% of serious accidents to 10% of people with COVID-19 are the source of 60% of new infections in two Indian states. But these are all different curves! And many factors would shift these: land justice in Italy, not counting bots in tweeter analysis, laws about accidents, variants of COVID-19 infecting people and under what economic and social distancing regimes.

Still, Gore is right that "we live in a largely Pareto world—inherently unfair, asymmetric and unpredictable" (Gore 2022). Yes, it is "unpleasant at first" to realize this, but crucial if we are to navigate a reality of black swans, unpredictable surprises as the first black swans they saw were to Europeans. Or long tails, the long low part(s) of power curves people who live to be over 100, like my beloved great-great aunt Gen, fall on in human age distribution graphs.

Dr. Robert Pearl argued in May, 2020 that we could use the "80/20 rule" to save the economy by sacrificing some lives. Since 80% of the deaths from COVID-19 are among the oldest 20% of the population, isolate them, he said. Make them take major precautions. Not horrible advice, but not great. To begin with, there is a false equivalency between the economy and human lives—they are not the same. As President Nana Addo Dankwa

Akufo-Addo of Ghana pointed out when ordering a lockdown, "we know what to do to bring our economy back to life. What we do not know how to do is bring people back to life" (Mackenzie 2020, p. 42). Dr. Pearl's conflation of the two leads him to argue for reopening the economy for the 80%. But we now know his strategy would have killed millions. His whole argument also ignored Long COVID.

While the 80/20 approach was never officially implemented, tightened protections and prioritizations for nursing homes, retirement communities, hospitals and the sick and elderly in general inevitably emerged. But the 80% was not abandoned, since COVID-19 turned out to be much more dangerous than Pearl thought. Lockdowns usually applied to everyone but essential workers, as did other quarantine measures. Treatments and vaccines eventually were widely available, with the expected disparities based on class, gender, color, and colonialism that maintain the viral underclass. Culture, politics, and technology shape every viral cycle.

The life of every virus follows the inevitable pattern of all such stories: birth/creation/origin, proliferation, mutation or sexual reproduction, individual death and group extinction. Because viruses are so small, viral lives are short; because their populations are so large, viral time can be quite long. Because they mutate incredibly quickly change can be fast, accelerating exponentially.

Humans live in a time scape that is constrained between the speed of our consciousness, the length of our attention, and the range of our understandings—so seconds to decades. But that range does not help us understand viral time. The microseconds of digital interactions or biological virus reproduction and the millennia involved in most biological evolution are beyond the consideration of most people, and necessarily are abstractions to those of us who do wonder about the Big Bang, the emergence of stars, the origins of life, deep history, and the end of time. Still, the harder we try the better.

Despite the incredible speeds at which viruses can propagate and mutate, the perspective of Virus, is very long term as well. The startling emergence of Long COVID in up to 20% or more of COVID-19 patients is a perfect example. Years after its discovery we are a long ways from understanding and therefore dealing with it. This is not surprising. Even though we've known for many years that many diseases caused by viruses leave people effected long after—measles-shingles, polio, HIV, and flu—this has been pretty much ignored. Other types of viruses can linger in the body politic, such as fascism. Thought eradicated in 1945, it is clearly still with us.

The 1918 influenza pandemic left many victims chronically ill, including Amelia Earhart with sinusitis that affected her sense of balance among other things. She lived around her infirmity, as most people with lingering disabilities or who reach old age did and still do, unless it contributed to the crash that killed her. Long flu was just not a thing, but it is real enough. Mental health admissions in Norway were seven times higher than average in the six years following the 1918 pandemic, although the two weren't connected until more than half a century later (Spinney 2017, pp. 218-219).

Long COVID is a good reality check that counters the natural human tendency to declare bad things over when they actually are not. Maybe if we had been thinking in terms of Long Fascism its current resurgence would have been less likely. A good lesson can be a horrible reality all the same. The toll of Long COVID, still sickening many people five years on, may eventually pass the cost of the initial pandemic.

Long COVID patients (long-haulers) suffer from a wide range of problems. There is brain fog, inflammations of almost all major bodily systems, heavy impacts on the autoimmune and nervous systems (autoimmune disorders and dysautonomia) and death. Some people have persistent viral infection in remote pockets of their body, others have systemic inflammation, with different syndromes producing the same symptoms. Long COVID is going to be sickening and killing people for decades.

Death is a necessary stage of every life cycle. Biological viruses go extinct, cultural viruses are forgotten, and digital viruses die when the machines and programs that host them become obsolescent. But all can survive through variants, produced biologically by mutations, culturally by creation, and digitally generated through programmed machine learning, copying errors and coders. Biological viruses can be engineered as well, modded or even hacked. Culture changes in the same way, through both unconscious and conscious processes, viral or not. As our understanding of biology grows, helped by viral algorithms, so does digital culture expand. The viral aspects of our lived reality, politically and ecologically, seems a growing fever dream, inevitably destined to grow exponentially to the point of transformation or collapse.

The final principle of the language Virus is how weird it is, how uncanny, how liminal. The viral is always on the edges, always "outside" of what we know in a very real way.

Language Is a Virus from Outside Space

Everything about the hawk is tuned and turned to hunt and kill...I discovered that when I suck air through my teeth and make a squeaking noise like an injured rabbit, all the tendons in her toes instantaneously contract, driving her talons into the glove with terrible, crushing force. This killing grip is an old, deep pattern in her brain, an innate response that hasn't yet found the stimulus meant to release it. Because other sounds provoke it: door hinges, squealing breaks, bicycles with unoiled wheels—and on the second afternoon, Joan Sutherland singing an aria on the radio. Ow. I laughed out loud at that. Stimulus: opera. Response: kill. But later these misapplied instincts stop being funny. At just past six o'clock a small, unhappy wail came from a pram outside the window. Straight away the hawk drove her talons into my glove, ratcheting up the pressure in savage, stabbing spasms. Kill. The baby cries. Kill kill kill.

—Helen Macdonald (2014, p. 83)

Essentially, the central provocation of the idea of "viral culture" is that we will need to come to terms with the problem of virulence, contagion, and the problem of the communication of what Baudrillard calls evil.

—Mark Featherstone and John Armitage (2020)

My first hard lockdown in Santa Cruz was not particularly dangerous. Infection rates were low, hospitalizations even lower, deaths rare. But we still felt it. We were in exile from our lives. Albert Camus writes about this in *The Plague*, how it felt before the impact of the epidemic that frames the story becomes clear in illness, death, and too many corpses to dispose of with dignity. Since the novel is also about the Nazi occupation of France, it echoes the "phony war" phase before Germany crushed France, and also, especially, the phony peace of the Vichy regime.

...the plague forced inactivity on them, limiting their movements to the same dull round inside the town, and throwing them, day after day on the illusive solace of their memories. For in their aimless walks they kept on coming back to the same streets and usually, owing to the smallness of the town, these were streets in which, in happier days, they had walked with those who now were absent.

Thus the first thing that plague brought to our town was exile...It was undoubtedly the feeling of exile—that sensation of a void within which never left us, that irrational longing to hark back to the past or else to speed up the arc of time, and those keen shafts of memory that stung like fire. (1975, p. 71)

I felt this, exiled in my own home. I didn't understand it though, until I read *The Plague*.

Before language developed in a small band of humans and infected the rest, it was not possible to worry about viruses and their principles at all clearly. All we had was feelings, gestures, touch. A language creates a particular model of reality but it isn't real. Or perhaps it is better to say it is only a slice of the reality that particularly interests humans. Even with the magic of story, poetry and song, words do turn back from many great truths. But language is the best tool we have to examine the types of viruses we need to understand most. Not just SARS-CoV-2 and its ilk, crucial as it is to know them to live with them, but also the viral ideas that often determine just what we can know, and therefore do.

William Burroughs (1967) warned us that "Language is a virus from outer space" in his *Nova* trilogy about mind control through mass media, sexual seduction, electronics, psychic interventions, and drugs. Laurie Anderson spread the news far and wide in her 1986 song of the same name. Language is all about mind reading and control, after all, and when it is used to manipulate us it certainly feels like an alien force that is both prosthesis and parasite.

When a parasite also benefits its host, adding to its function prosthetically, it is a symbiont. All organisms are symbionts originally, because life started as an alliance of almost-alive systems. Since language comes from outside us it is a parasite, a symbiont that intimately links us into the vast super organism that is human culture, greatly increasing our ability to expand outward, to extend the human throughout the biosphere—not just our bodies but also our micro-plastics, the heat of us and our works manifested in global warming, and our ravenous hungers producing mass domestications and extinctions.

So maybe speaking Virus is a way to domesticate the viral, even learn to thrive with it. We can hope. Right now the viral is almost always beyond our control, we can't even count how many different biological viruses there are. There is progress. We now know they originated from ancient cells back at the very beginnings of life even if humans only discovered and named them recently (Nasir and Caetano-Anolles 2015).

During the influenza pandemic of 1918 a number of doctor-scientists realized it was spread by an infectious agent. René Dujarric de la Rivière named it *virus*, although he didn't know if it lived or not. It took a great deal of searching in the 20th century to capture and study actual viruses; they were too small for existing filtration devices. But scientists persisted, and thanks to a ferret sneezing in the face of a researcher named Wilson Smith, who then caught the flu, they zeroed in on the problem and showed it was either an organism or a toxin that spread it. By the 1950s, it was clear it was biological but how it reproduced was still a mystery (Spinney 2017, p. 181).

It turns out viruses reproduce and change in a number of ways. Each replication introduces errors and some of them successfully proliferate, thanks to luck or better fitness. This is called "antigenic *drift*." It is supplemented by viruses swapping genes, in a process called "antigenic *shift*," also termed reassortment or, more fun but less accurate, "viral sex." When the viruses merging are originally hosted in two different species (pigs and humans, birds, and pigs) or more (pigs, birds, humans) you can get decidedly new viruses that might elude existing immune responses. This is what happened with influenza A, where "shifts" led to H1 in 1918, H2 in 1957, and H3 in 1968 (Spinney 2017, pp. 185–187). Mike Davis compared the impact of antigenic shifts, and their reassortment of specie-specific "keys" to open up target cells, to revolutionary change, as opposed to the "reform" of antigeneic drift (Davis 2020, p. 5).

The bottom line is that while complex creatures such as ourselves might take 8 million years to change our genetics by 1%, an RNA virus can shift that much in a matter of days (Davis 2020, p. 3).

The weirdness of biological viruses is revealing. Almost all life uses the same four bases—adenine, guanine, thymine, and cytosine—in their DNA and RNA. But some viruses, over 200 phages that eat bacteria, use 2-aminoadenine instead of adenine, a more stable configuration (Capelewicz 2021c).

Almost all creatures are saturated with viral symbionts. The herpes that infects so many of us was apparently first spread during bronze age kissing (Sample 2022). In 2022, a pig virus killed a pig heart that had been transplanted into a human man, killing him. A similar problem has cropped up with pig hearts transplanted into baboons (Yang

2022). We have Long COVID because the SARS-CoV-2 lingers in some bodies in hearts, lungs, brains, and other important tissue. Even when it doesn't, sometimes it has unleashed newly discovered subcellular "agents" in our cells called "inflammasomes." They trigger caspase-1 and gasdermin D. This, in turn, burns up the infected cell through pyroptosis ("fiery death"), the process of running the cell into fatal inflammation (Goodman 2022b).

"Surprisingly, the male reproductive tract lit up like a Christmas tree" is not as jolly as it sounds. A researcher is explaining how they found out COVID-19 infects at least four parts of the male genital tract in nonhuman primates (McClure 2022). These inflammations lead to performance issues such as COVID Dick, as the comedians have charmingly named it.

It isn't just primates that get COVID. Bats, minks, zoo lions, household pets, and many other creatures can be infected. An Omicron-like variant has emerged in North American deer. While not easily transmissible to humans, this kind of wild animal reservoir of the virus is why it is unlikely to be wiped out soon, if ever (Goodman 2022a).

One form of weirdness that occurs in all forms of the viral is zombification. Biological viruses go so dormant while they wait for a host to come along they are considered not even alive, making them zombie viruses called virons. While never fully eradicated (99% gone but still in Pakistan, Afghanistan, and, now, Rockland, New York), polio seemed extinct but is making a zombie-like comeback (Madad 2022). The 1918 influenza virus has been brought back from extinction at least twice (Spinney 2017, p. 187).

Pop culture echoes this. There are at least 150 movies about viral outbreaks. Some are quite realistic (scary), such as *Contagion* (2011), which eerily describes the COVID-19 pandemic if it was ten times more deadly. A surprisingly large number are about viruses that produce zombies or zombie-like behavior, leading to great disappointment among some nerds. Their "I Wanted Zombies This Virus Sucks" slogan went viral. But it is a whistling in the dark kind of joke, feeding off the fear of the unknown that permeates society today, buffeted as it is by viral forces.

Biological contagions, usually viral, turn people into nothing more than ways of spreading the virus. Digital contagions, always viral, transforms recruited computers into zombie networks that not only spread the virus but also can attack other networks with ransomware and other malware. Culturally, the most potent viral ideas can make people insane, shrinking their lives down to trying to spread the wisdom of whatever cult, irrational conspiracy, or mad fantasy that has colonized their brain.

In the digital realm, the spread of viral AI algorithms that drive roboprocesses has been described in detail as zombification. The anthropologist Catherine Besteman argues that it results in "deskilling and disciplining" workers. She goes on to warn,

Roboprocesses also spur the zombification of interpersonal relationships—the robotic click of the Like button as a replacement for discussion and debate, the relentless barrage of commercial pop-up ads that track networks among internet users so as to turn personal relationships into marketing opportunities, the use of programmable robots as companions for the elderly who are abandoned in nursing homes by other living people. (2019, p. 174)

She concludes that "Zombification by algorithm is what happens when algorithms program policy and structure social interaction rather than the other way around." This is not a new phenomenon. John Quiggin published his *Zombie Economics: How Dead Ideas Still Walk Among Us* in 2010. Economists have long known that trickle-down (tax cuts for the rich) is a lie, for example, but the idea is just too useful for vampire capitalists to let it fully die. Zombie thinking is increasing. Besteman quotes Paul Krugman warning of "zombie lies" and "zombie ideas" producing a "Zombification [that] degrades critical thinking and the ability to imagine alternatives" (p. 175).

Economic zombie arguments drove debates over COVID health measures such as lockdowns, masks, and quarantines, claiming such measures should not be taken because they would hurt the economy too much. As if millions of dead and hundreds of millions sick wouldn't hurt the economy. As if much of what passes for "the economy" is the unnecessary—but oh so profitable—buying and selling of useless things and people's irreplaceable time.

In the summer of 2021, thanks to information spread by Algorithmic Intelligence-driven social media such as Twitter, Instagram, and Facebook, 30% of Republicans thought COVID-19 vaccines included microchips from Bill Gates, allowing him to turn all recipients into cyborg zombies

(Devega 2021b). The totally different, but equally inane, theory that the RNA COVID-19 vaccines were actually "operating systems" to produce more SARS-CoV-2 and make all recipients a different kind of zombie, also spread digitally around the world (Fauzia 2021).

Viruses are a type of zombie. They exist only by infecting living or digital or cultural hosts with their replication machinery, turning them into an army that needs one thing: more human brains to tell about QAnon, or cells to make SARS-CoV-2, or AI-generated digital variants to try out on Facebook.

This proliferation of the zombie trope is a sure sign of the uncanny. Humans are social animals with brains evolved to pay special attention to human faces and their changing expressions. Other humans do offer us the greatest joys and the greatest dangers. The uncanny is the tingling mix of anticipation, pleasure, and fear we get when something triggers our human recognition modules and yet doesn't feel human. This gulf is the uncanny valley. Zombies, cyborgs, androids, vampires all live there and feed on the fears and desires the paradox of the uncanny generates. It makes your skin crawl, doesn't it? But maybe in a good way. Many monsters are hopeful.

To be human is to feel uncanny. We are at home and not at home with our bodies. Because of the lens of consciousness our very nature is familiar and yet uncomfortably strange. Our models of reality are always in tension with reality itself, which is too complex to predict consistently anyway. Even that sliver of reality we have evolved to interact most with, the vulgar physics of Newton, of hunting and hunted at the water hole, is truly beyond our ken. Base desires, the blood and mucus and excrement of living and the inevitable totality of death, are concealed by our manners and religions, yet they are always present. That itch in the mind that cannot be soothed is always there.

Even if we grow comfortable in the world into which we are born, that world is soon gone, swept away by relentless and accelerating techno-scientific-cultural change. Our first technologies came directly from wild nature, as we did, and were intimate with it: fire, wood, stone, plants, and domesticated animals, starting with ourselves. Today we transform everything and make our Earth uncanny. We feel losses we do not know how to articulate.

It gets weirder—the spiraling of our fates, the intertwining of our destinies, with biological/digital/cognitive virality. This is why speaking Virus is particularly important now, for processing—living with—how quickly our world is shifting around us. We have the grammar now, but that is only the skeleton of a language—vocabulary is its flesh.

2

Viral Vocabulary

'Ronaverse Slang

What slang really does is show us at our most human.

—Jonathan Green "Mister Slang" (Steinmetz 2018)

My youngest son, Z, was traveling in the far south when the pandemic caught up with him in El Calafate, Patagonia. He was there to see the penguins when Argentina implemented a draconian lockdown. He found a little apartment with good internet so he could do his classes at the University of Montana remotely. He was there more than four months, more than three quarantines. Quarantine from the Italian quaranta giorni, "40 days"—the isolation period demanded by ancient ports. For COVID-19, should it be quattrocentine (400 days) or ottocentatine? Many of the words we used for the COVID pandemic have old roots, revealing something of what humanity has learned dealing with deadly contagions.

Plague, Latin *plaga*, a wound or a stroke, probably from the Greek *plangere*, lament by beating the breast, related to *plaga*, a blow, from the Proto-Indo-European root *plak*—to strike. So the ten plagues of the old testament aren't all diseases, they are blows from God, including rains of frogs and rivers of blood. Perhaps the Old Irish *plag* first focused on pestilence. By the Middle Ages that meaning was common, and applied often to bubonic plague, as in the German *Plage* and the Dutch *plaage*—contagions that caused many deaths. Old meanings are not lost. For many, COVID-19 is indeed a plague from God, and for even more people it has been a blow leading to lamentations. Even those of us who have yet to get infected feel pummeled.

Or consider host. It has at least three origins: (1) Latin *hostia*, sacrifice, became the consecrated host of Catholics; (2) Medieval Latin give us *hostis*, stranger or foreigner, in classical Latin an enemy, in English today an invading army; and (3) oldest perhaps is *ghos-pot*, "guest-master" from Proto-Indo-European, as in the computer that holds programs or the animal infected by a virus. Viruses are parasites and need hosts to live, let alone thrive. Hosts are also growth mediums. Biological viruses need cells, digital viruses computer programs, and cultural viruses human minds. Cultural viruses don't just survive in living minds, they can lie dormant, cultural virons, in books, movies, songs, and fashion. But they do need minds to host reproduction and so mutation. So host's echoes of enemy, of stranger danger, of invading legions and of sacrifice remain.

Most of the key COVID-19 pandemic vocabulary already existed in virology, information science, politics, sociology, and cultural anthropology. But when technical terms go viral there is still the shock of the new as they morph. The crowning achievement of this epidemic of neologisms is the naming of the variants, a scientific-cultural-literary process described below. In other cases old terms can go through fascinating transformations. Bubble, for example, goes from a floating bubble to being clueless ("living in a bubble") to a place we were all supposed to be, safely socially distanced in lockdown (a prison term), in isolation, in quarantine.

Essential workers suddenly includes garbage collectors, Amazon drivers, hospital janitors, and gig drivers delivering food. Teachers, nurses, postal workers, and many more blue, pink, and white collar people are now lauded as heroes. The reality of who makes society work and its system of rewards and punishments—wealth to the few, a short life to the viral underclass—was suddenly clear. How essential, really, are oligarchs, or even hedge traders? The *Coronapause* has been revealing.

Yes, now we have *'Ronaverse* slang, often shared worldwide. In the U.S. it seems to have started with *the 'rona* and *Miss Rona*, quickly followed by *bubble, covidiot, maskholes, maskulinity* (the illusion that not wearing a mask makes one seem manly instead of stupid, coined by my son C), *maskne* (mask acne), *zoombombing, antivaxxers, vaxholes* (flaunting their vaccination status), *vactivists*. For getting *vaxed, jab* has invaded the U.S. from the U.K., but *Fauci ouchie* hasn't made it there. Working from home, shopping from home, recreating at home (*quarantinis*!) made every day seem like *Blursday*, the only day of the *Groundhog Pandemic*. We watched grimly for the next *scariant* or *Frankenvariant*. As time passes, *novids*, the never infected, became rarer.

Londoners deployed their cockney rhyming slang, so *Miley Cyrus* is corona virus and one catches *the Miley*; a person acting out of irrational fear is a *morona* and *coronalusional*. Inevitably it is *Covexit* for the end that may never come and certainly won't be what was hoped for if it does. In Canada and India, *caremongering* spread to counter scaremongering.

Australians have a mania for shortening words, so *quiz* for quarantine, *sanny* for sanitizer, *iso* for isolation and *pozzie* for positive. Unlike the Dutch and Germans who use *hamsters* for hoarders, the Aussies say *magpies*. And who can't enjoy the German *Öffnungsdiskussionsorgien* ("orgies of discussion") for endless debates about when to reopen, *coronaspeck* (corona-inspired stress eating), *Klopapierhamster* ("toilet paper hamsters" for hoarders) and "face condoms" for masks: *Gesichtskondom*?

The Spanish use *COVID* from the English. They also have coined *confinamiento* (lockdown), *cuarentenear* (to be quarantined), *mutearse* (to mute yourself), *infodemia* (infodemics), *zoomleaños* (birthdays celebrated virtually), *covidiota*, and a word invented in 1932 to describe spreading fascism that has made a comeback: *fascistoide*.

In China, the government spread *da bói* (大白, big whites, from the white robots in Disney's movie *Big Hero 6* (Hall and Williams 2014) for PPE white-suited COVID cleaners) but during the harsh Shanghai lockdown many started calling them "white guards," a reference to the Red Guards who terrorized the country during the Cultural Revolution or "Xi's Big White Army." Everyone seemed united in lauding *nìxíngzhě* (逆行者, going against the flow-sters, literally "countermarch heroes"), the frontline health care workers risking their lives by confronting the pandemic, not hiding from it.

Many of the new coronaverse words in China are to get around the heavy government censorship. So Wuhan become "*wh*" and Hubai (the province it is in) *Hu*. Criticism of the regional leaders where the pandemic started can't use their names or titles, so the governor of Hubai, mayor and Communist Party secretaries of Wuhan became *F4*, a popular Taiwanese boy band of the early 2000s.

Dr. Li Wenliang was a whistleblower in Wuhan when the outbreak started. His January 3, 2020 police interview was leaked. The police demanded he stop his "illegal activities" of warning about the virus. "Can you do this?" He responded, "Can." "Do you understand?" they asked. "Understand," Li replied. So "Can you do this? Can. Do you understand? Understand." went viral. Once that was censored, it became "I cannot and do not understand." Dr. Wenliang, aged 33, died February 7, 2020 of COVID-19. Posthumously, the Wuhan police apologized, and he was awarded a medal by the Chinese government.

Not that American racists care about such heroism. For them it was *Wu flu*, *Chicom flu*, and Trump's favorite, *Chinese Virus*. But in India it has been blamed on Muslims as a *Thook Jihad* ("thook" is spit in Hindi).

When it became clear that COVID-19 was most dangerous to older generations we heard *boomer doomer*, *boomer remover*, *senior deleter*, and *geriatric dispatcher*. This generational framing goes back at least to *limpa-velhos*, the "cleaner of the old" in Portuguese from Rio de Janeiro's experience in 1918 with the influenza pandemic (Spinney 2017, p. 53).

My favorite coronaverse terms were from Paul Ford (2022), who proposed *demic* for the different *panic* waves and *prepanic* and *panic* time frames. His vocabulary is a noble attempt to take viral time into account, at least how humans experience it. For example, he advocated replacing weekend with *binge* (Friday night to Sunday morning *good-binge*, Sunday morning to Monday morning *sad-binge*), *lightmode* and *darkmode* for night and day, *fade* for twilight, *fleshings* for minutes, and many more. So one could say, "I miss going to Europe *prepanic*. I hope to get back between the Omicron *demic* and Omicron-plus *demics*. Today I'm going to *sad-binge* until the *fade*. Don't know what, *Man in the High Castle* or *Saints & Sinners*. I'll decide at the last *freshing*." Or "Fuck, this new COVID *demic*, the resurgent Xec ("Zek") phase of the *panic* is hitting everyone!"

Less amusing are the technical terms that have escaped into the general discourse. Some are from biology, especially virology, others originate in the digital technosciences and cultural studies. The most important are defined below. Variants and their naming conventions are detailed after that. Finally, four crucial viral concepts—Hosts, Vectors, Catalysts and Controls—are given their own section.

Technical Dictionary

Virtues, like viruses, have their seasons of contagion. When catastrophe strikes, generosity spikes like a fever. Courage spreads in the face of tyranny.

—Nancy Gibbs (2009)

Protocol is how control exists after distribution achieves hegemony as a formal diagram. It is etiquette for autonomous agents. It is the chivalry *of the object.*

—Alexander Galloway (2004, p. 75)

Note how in this Virus vocabulary each word maps onto all domains of the viral.

Affordances This design term for features that "afford"—open up—certain modes of use is common in the digital world and deserves to become widespread. Going from the way things are to what they can become, fostering the right affordances in all forms of tech, from digital to bio to social, is crucial.

Attack Rate After a pandemic is over you can calculate this by taking the total number of cases and divide it by the population. Since later variants can reinfect people who have had earlier variants, this rate can be above 100%.

It can usefully be applied to other biological epidemics, as well as computer attacks (how many systems at risk are compromised) and the spread in culture of specific ideas (convincing what percentage of people exposed).

Case Rate (CR) The number of reported infection cases. This measurement is often deceptive. It all depends on how the cases are defined, collected, and reported—and by whom.

Case Fatality Rate (CFR) Deaths as a percentage of cases. This measurement is only as good as the accuracy of the CR. With a virus such as SARS-CoV-2 or with contagious ideas or computers exposed to viral processes, one doesn't always know the number of people or machines infected that don't develop symptoms or infections that are not recorded. Closed systems (cruise ships, cults, corporate networks) can easily generate useful data. For open systems large scale testing is necessary.

Community Spread Complex contagions infect best through the strong links of community.

Contact Tracing Tracking digital, cultural, and biological viruses is often crucial but sometimes irrelevant. If contagion spreads quickly, or if it is hard to test for, contract tracing cannot help contain it.

Singapore and China did well with COVID-19 contact tracing at first. Singapore had 5,000 contact tracers for a population of 5 million. In the U.S., an equivalent would have been 330,000 tracers (Christakis 2020, pp. 111-112). Since the Center for Disease Control's (CDC) first mass produced test was flawed, and many leaders were in denial about COVID-19, contact tracing was never going to work in the U.S.

Contagion The spread of a disease, powerful idea, emotion, or malware through a system.

Correlation Does not prove causality but it is interesting evidence. If events or ideas are highly correlated, it could be one helps cause the other, or something causes them both, or perhaps there is no relation.

Consider these fallacious correlations: the relationship between Nicolas Cage films being released and deaths by falling in a pool, per capita cheese consumption and people dying from becoming entangled in bed sheets, and the close linking of the marriage rate in Kentucky and the number of people who drowned after falling off a fishing boat. These types of correlations QAnon calls evidence.

Excess Mortality Comparing the deaths before a pandemic to during and after is the best way to measure its actual toll. If other disasters or new forms of tracking deaths don't confuse the data, it captures not just direct deaths, but the knock-on effects of limited health care and the positive results of masks and other isolation policies.

Flattening the Curve Taking public health measures (quarantines, lockdowns, masks, and vaccinations) to lower the rates of infection to a point that won't collapse the health system.

Endemic A disease, algorithm, or idea typically found in limited numbers in a group and/or place. If unlimited in a defined area, it is epidemic. If unlimited globally, it is a pandemic.

Epidemic A disease, algorithm, or idea spreading exponentially in a defined area. If it is unlimited globally it is a pandemic. If it is limited in numbers to one group and/or place, it is endemic.

Exploits Flaws in systems that afford changes, be they malicious or beneficial.

Generation Interval As important as the reproduction (R) rate. Eight cases after four weeks could be R of 8 or R of 2, depending on the generational rate, which is how quickly the infection passes on. This rate changes for many reasons, so R is never stable (Cepelewicz 2021b).

Genetic Susceptibility Viruses and their hosts have varying inherited weaknesses. Some computer systems and some people are more likely to get infected than others, depending on their configuration, history, and exposure.

Herd Immunity This is achieved when the reproduction rate (R) is below 1, so the virus population is declining. That doesn't mean it won't survive. It could become endemic, it could lurk as virons, or it could mutate and as a variant start thriving again.

Because COVID-19 mutated so quickly, old models were unhelpful. The level of infection or vaccination needed (20% the optimists said, 60% was the conventional wisdom) never became clear. That the Omicrons easily spread to those who had earlier infections was a shock to many experts. Overall, most theorizing about herd immunity ignored the danger and reality of variants, although variants is what makes flu a yearly trial. Viruses are moving targets because viruses are always evolving (Harnett 2020).

Index Case The first person or machine in a group infected with a virus.

Infection Occurs when new information (biological, digital, or cultural) modifies a host.

Infection Fatality Rate (IFR) As with the CFR, the IFR is hard to get right. Even closed systems, such as the *Diamond Princess* cruise ship, produced a wide range of estimates because of a plethora of different errors (Shen et al 2021).

Links Can be either strong or weak. Weak links are not necessarily less effective conveyors of infection. For some contagions they are more effective overall than strong links. Networks of weak links are called small-world topologies, as in the six degrees or less of separation between you and Kevin Bacon. Not only can he be linked to any another actor with just a few steps, it is almost certain that if you could track it, you know someone who knows someone who knows someone who knows someone who knows someone who knows Kevin Bacon. This is because the network is exponential—each one of your acquaintances has dozens of their own and theirs have dozens more. Put scientifically, "the distance between any two people is on the same scale as the *logarithm* of the size of the network. Thus, the degrees of separation between people stays incredibly small even as the population size becomes vast" (Centola 2018, p. 26).

Strong links are between family, friends, and collaborators, or in the case of machines between those in one small dedicated network. But you can also make strong links through specific actions, such as sweaty dancing in a small club or inserting a thumb drive in your PC. Every person or machine is a node in network architectures, and they almost always have both strong and weak links. Weak links between highly connected hubs (a node or group of nodes) are a scale-free network, like the one that has spread SARS-CoV-2 and its variants around the world (Centola 2018, p. 19).

Mutants All viruses are mutants. All viral reproduction is by gene mixing or copying. When errors occur mutations arise. The environment influences mutations through epigenetics. When a toxic chemical or a cosmic ray hits a virus changing its DNA or RNA and it reproduces, this is a mutation. If it survives, Debora Mackenzie points out, "That is not random. That's evolution" (2020, p. 135).

Networks Without connection contagion is impossible. The pattern of connections is a network. Some networks are more conducive to contagion than others. Assortative and disassortative networks are different. Assortative networks are of highly connected nodes (people, machines) that are linked mostly to other highly connected nodes. Outbreaks spread quickly through the core of high-risk nodes but fail to go beyond its limits. Disassortative networks have highly connected nodes linked to less connected. At first contagion spreads slowly, but then it accelerates through the many weak links, leading to a larger epidemic in the long run (Kucharski 2020). Viruses propagate successfully in both configurations.

Pandemic An unwanted disease, algorithm, or idea spreading rapidly globally. If confined to one area it is epidemic. If it is limited but ongoing, it is endemic.

Patient Zero The original. The viral AI "run" or launch, the first infected human, the culture maker (artist/writer/rapper…). Tracing Patient Zero is similar across domains. Charting small changes in genetics, in code, in story details over time can be used to estimate lineage and age.

Often, Patient Zero is misattributed, as with AIDS Patient Zero who wasn't, yet he was vilified. HIV was infecting people decades before the pandemic broke out. He was, however, a super-spreader (Thrasher 2022, pp. 51–55).

Percentages If something happens to us or someone we know it is hard to remember that percentages rule viral phenomena. Catch COVID-19 and it is mild, have a friend get sick after being vaccinated, and it is easy to ignore that these events are only one point of data, and not proof that the SARS-CoV-2 is not deadly or that vaccines don't work. Do you know that 8% of 25 is the same number as 25% of 8? Does it still confound you that there is a 50% probability that in a room of 23 people two have the same birthday? There are many paradoxes that come from the difficulty humans have understanding percentages and yet, to speak Virus, we must master them.

Personal Protective Equipment (PPE) One can buy all sorts of hardware and software for personal digital or biological protection, to go with the practices (long passwords, wearing a mask in sweaty crowds) that can lower the chances of infection. Education, diverse networks of friends, families that aren't all crazy, can help protect against deleterious, as opposed to helpful and/or fun, cultural viruses. But the cultural realm is much less stable than the biological and digital, so cultural PPE is way more complicated.

Phages These are viruses that feast on bacteria, even antibiotic resistant versions. Digitally, we could think of a phage as anti-malware viral software: culturally, it would be some individual or group that uses viral processes to combat viral conspiracies.

Protocol Treatments for infections. The coding rules and conventions for computer systems. The written and unwritten conventions of diplomacy. The policing of typical human interactions, such as dining together. Viral processes provoke similar protocols across domains.

Focusing on digital and political systems, Alexander Galloway (2004, p. 242) has argued that protocol can help solve "the problem of hierarchy" and should guide our efforts. Good advice, as long as we remember protocol is a process, not an algorithm.

Reproduction Rate (R) An average of how many successful takeovers—the reproduction of the virus, algorithm, or idea—each infection produces. R is a crucial measure for setting health, political, and digital policies. R constantly changes because transmission efficacy, infectiousness, and the contagion itself can change rapidly—through viral mutation or social responses, for example—and they are often hard to measure in any case.

Ronald Ross set the first reproduction rate, on malaria. Working for the Indian Medical Service he developed his "mosquito theorem" by calculating that if one infection results from every 1,000 mosquitos then lowering the number of mosquitos could make the epidemic die out. He argued that only 1 in 4 malaria carrying *Anopheles* mosquitos ever bites someone, so 48,000=12,000 bites. Only 12 of those 12,000 biters are effective transmitters, he went on. Mosquito life spans are brief, so he assumed only 1 in 3 (4 of the 12) would survive long enough to become infectious, and only one of them would succeed in biting a human and infecting them. Since 20% of those infected recover each month, the rate of infection has to be greater than recovery.

This means that one doesn't have to get rid of all mosquitos, just kill or hide from them. Sleeping nets are enough to drop the reproductive rate below one. Infection is a "dependent happening" in Ross's schema, because what happens to individual mosquitos and people depends on what has happened to other individuals (Kucharski 2020, pp. 17-18).

R was formalized and simplified by the mathematician Klaus Dietz, who proposed it for the overall measure of a disease spread. Ebola and pandemic flu have Rs of 1-2; SARS 2-3, Smallpox 4-6 and chicken pox 6-8. Measles can have an R of 20 or more (Kucharski 2020, pp. 54-55).

R varies by target population, spread modalities, and context. Different variants could well have different Rs. Omicron variants spread much more quickly than earlier SARS-CoV-2 types. R is different in different times and spaces. It is a situated knowledge.

Shelter-in-Place A nicer way of saying lockdown.

Social Distancing A nicer way of saying social isolation through physical spacing. It is also air gaping computers and fire walls, social media fasts, and other refusals.

Super-Spreaders Can be creatures, machines, or events. They follow the Pareto Principle. The majority of infections of all types are caused by a minority of spreaders.

The Election Integrity Partnership of academics analyzed "election misinformation" on social media (Facebook, Twitter, Instagram, YouTube, and TikTok) several months before the January 6, 2021 Capitol attack and found that a handful of posters were super-spreaders "responsible for the most frequented and most impactful misinformation campaigns."

They were Trump, his two elder sons, and other paid and unpaid far-right activists. Trump spread six specific sets of misinformation that were central to the worldview and actions of the rioters (Paul 2021).

Testing Knowing what is spreading and how far is crucial for control. Without accurate testing many responses to pandemics are not possible. Successful testing that shows the extent of a contagion is a problem for some, so Trump claimed that less testing would mean fewer people infected.

No test is 100% accurate. There are false positives and negatives, as variations change test effectiveness declines, and the level of testing varies considerably. Less intrusive and more inclusive testing regimes can be very helpful, as in testing sewage for signs of antibodies to SARS-CoV-2.

Variants The viral life cycle relies on the creation of countless varied progeny through combinations and mutations. Daughter viruses that differ significantly from the original and thrive are variants. This is a good example of why perfection is not always ideal. If viruses only copied themselves perfectly they would not survive.

Virions When biological viruses are not in a host they are virions, enclosed in a protein shell (capsid) for moving between hosts or just waiting for one. They seem significantly less alive than viruses. A virion digitally or culturally is infectious information waiting for access to a working computer or a human brain, to occupy and modify so as to reproduce itself.

Rhetorical Variants

The only road from grammar to logic, then, runs through the intermediate territory of rhetoric.

—Northrop Frye (Gray 1997, p. 96)

We live in and through the act of discourse.

—George Steiner (Gray 1997, p. 182)

I don't think I've ever feared any disease as much as I did COVID-19. In Istanbul in 1972 I had a very bad night with a nasty fever. I had hallucinations; I almost went insane. In the large dormitory of cheap bunk beds four levels high two people died that night, Japanese travelers as poor as I was.

But my fever broke, some wild Canadian nurses bought me some powerful antibiotics, and I never was quite that sick again, although I had a bad cough for the next six months. In Barcelona in 1979–1980 I had an infected tooth that gave me significant trouble until it was lanced. That New Year's Eve I was almost as sick as I had been in Istanbul and thought I might die. Since then I've had some bad bouts of flu and violent traveler's upsets in Turkey, Morocco, Egypt, and Mexico, but I was never afraid. That's life. But now that I'm coming to the end of my days the possibility of getting Long COVID and losing much of my energy, or worse, my ability to think, brought me to a new place, a fearful place. Fortunately, I don't do fear for long, I get bored too easily. But when fear impacts my behavior significantly, I notice.

At the start of the pandemic I was hyper vigilant and each variant seemed significant. A few months after they appeared most became history. Few achieved long-term dominance. Past variants are important because they can tell us what might come next. We track them through their names, but that is often confusing because of the complicated system that names viruses and the diseases they cause.

The naming of things is an important part of rhetoric, as is the study of using language to exercise power: explanatory, seductive, coercive. Labels are a very important form of this, revealing both who has the power to name and what the right names can do. We know this from the long traditions of magic and politics, and the new digital realm where passwords (your secret names) are the gatekeepers to knowledges and their powers.

COVID-19 is revealing—"COronaVIrusDisease-2019." COVI because it is a coronavirus, with a distinctive crown-like feature. SARS-CoV-2 (named by the Coronavirus Study Groups, part of the Committee on Taxonomy of Viruses which is the International scientific group that categorizes viruses) since it is a close genetic cousin to SARS (severe acute repository syndrome).

SARS-CoV-1 killed almost 800 people out of 8,000 infections. In 2004 it spread from Asian palm civets to horseshoe bats to humans in Xiyang Yi Township, Yunan, China. It was contained but seriously scared public health professionals around the world. SARS-CoV-2 is actually a variant of this virus, much more infectious, much less deadly. This is a good reminder that variants are not synonyms. They represent viral semantic change, also known as a semantic shift, a lexical change, a semantic progression.

The naming of virus variants has only recently been formalized in the Pango system. It was established by young researchers, starting with Áine O'Toole, a doctoral student at the University of Edinburgh. She coauthored a paper in April, 2020 that put forward a naming procedure of letters and numbers based on the genealogy of the viruses. She began applying this nomenclature to SARS-CoV-2. Emily Scher, a postdoc at the same lab, wrote a machine-learning algorithm to speed up the classifications. Together, they named the software "Pangolin" after the odd scaly anteater once suspected of being the source of SARS-CoV-2, which was soon shortened to Pango (Ferguson 2021).

When WHO first declared the COVID-19 pandemic in March of 2020, the GISAID (the Global Initiative on Sharing All Influenza Data) database had 524 sequences. A month later it was 6,000. In two months 35,000. By the end of June O'Toole had sorted 57,000 sequences into 39 variants. In November, as she was finishing her dissertation, she did her last solo sorting—200,000 sequences (Ferguson 2021).

After that, the work became more collective and a number of other researchers joined her. There is now an international Pango Network Lineage Designation Committee which collects the different virus sequences and sorts them into lines of descent. In the spring of 2021, their Pangothon sorted over 800,000 sequences into roughly 1,200 lineages. By that summer there were 2.5 million SARS-CoV-2 sequences in the data bank (Ferguson 2021). At the end of 2024 there were almost 20 million (Ma et al. 2024).

The Greek names for variants come from WHO. They felt the scientific terminology wasn't easy for the media and public to understand. In late 2020 WHO started breaking down the most potentially dangerous variants into those that are being Monitored, of Interest, of Concern, or the worst—of High Consequence. By the end of the year, there were 14 variants being monitored, two of interest and four of concern—Alpha, Delta, Gamma, and Beta. Even without the baroque labels, any variant that gets a Greek letter seems scary.

The Greek letter system has its problems. For one thing some letters can't be used as they are confusing, Xi is the first name of China's ruler and Nu sounds like "new." Besides, Alpha is not the original form of SARS-CoV-2 and 2024's Xec ("Zek") is considered a subvariant, of Omicron, so wasn't given a Greek letter.

In the fall of 2021 many experts predicted the Delta surge would be the last wave in the pandemic (Karlis 2021b). Turns out, not so much. But Omicron seems to have been. Not because it really is the end of COVID-19, but by generating multiple variations that spread easily through defeating vaccines and other immunizing factors, but with milder symptoms, it seems to have been the variant that ushered us from pandemic to epidemic to endemic.

That doesn't mean that it isn't a killer. Although most Omicron versions are 50% or more less deadly than earlier variants, some are also 50% more infectious. Mathematically, this is significantly more deadly than a 50% higher IFR (Resnick 2021a). The number that die is a percentage of the numbers that get it. Make the second number grow and fatalities rise exponentially. Make the first number grow and the death rate goes up arithmetically. The same math applies to Long COVID. Even if the fatality rate is half of Delta (5% instead of 10%), if twice as many people get Omicron you have the same number of deaths (Antonelli et. al. 2022). Fortunately, Xec isn't that deadly.

I spent Thanksgiving 2021 alone. My son C was with his girlfriend's family and J was with hers. I had lots of time to watch football and doom scroll the Interweb. Among the dooms I read about that day was a new variant out of Botswana B.1.1.529. It was named Omicron the next day by WHO.

Omicron was surprising because it isn't descended from Alpha or Delta. It evolved from a common ancestor from the first few months of the pandemic, "practically a geologic era in viral-replication time" (McKenna 2022). The main origin theories are an animal source, "cryptic" spread where there isn't good testing, or long-term infection of someone immunocompromised, perhaps with an untreated HIV infection (Nurith 2021). This last hypothesis seems likely, considering the ornate architecture of Omicron and its origins near South Africa, perhaps Botswana. Ironically, the failure of the world to fully deal with the HIV epidemic in Africa could well have led to Omicron emerging and killing millions who survived earlier waves. But other scientists argue that it had been evolving in mice and then "spilled back" into humans. In experiments they showed the Omicron spike was a better fit for infecting mice cells than human (McKenna 2022).

Whatever its origins, Omicron was also surprising because it has twice the mutations as Delta had on the spike protein, producing significantly more infectious. It showed powerful "antibody escape," and its variants were less deterred by earlier infections or vaccinations than earlier manifestations.

A team of *Atlantic* staff writers have documented how the response to Omicron reprised the same errors that had been made with every major variant in the first two years of the pandemic: it is labeled "mild" and therefore less of a threat, vaccines are assumed to be perfect shields, testing is heralded as 100% accurate and dependable, the pace and scale of the variant's spread is profoundly underestimated, not protecting the most vulnerable groups, and laying off most of the responsibility to dealing with it to health-care workers. Omicron went from 13% to 73% of the cases in the U.S. in a single week when it first hit (Wu, Yong, and Zhang 2021).

With variants, there are no guarantees that they will be milder. They can well be worse. This is as true of viral variants that occur in digital and cultural domains as biological. QAnon was not the first Anon, it was proceeded by postings from CIA Anon, CIA Intern, and WH Insider Anon among others. It was just the variant that proved extra infectious and virulent. It might have come out of the churn of Reddit conspiracy culture, where imitation is indeed the highest form of flattery, but it was probably a planned test. Live test marketing has been integral to advertising and military cyber operations for decades. With digital iterations cheap and fast, all social media marketers, PSYOP operatives, culture vultures, reality hackers, and cyber-revolutionaries have engineered cultural viruses in their toolkit.

While variations can arise out of natural dynamics in both cultural and digital modes, they also can be programmed consciously. Facebook, Google, Amazon, and the other social media behemoths generate countless minor variations for all their profit-driving algorithms, and then implement whichever does best, even if it fosters a fair amount of madness. All these leviathans, and indeed all mass market corporations, do the same, as do governments engaging in hybrid war. When China spreads their own or Russian disinformation virally, as it does, they change it in some ways, turning out different versions just as a biological contagion generates variants (Hvistendahl and Kovalev 2022). Then they propagate the most successful.

The similarities in terms of vitality across domains is particularly visible when looking at four of the most important words in the Virus vocabulary: Hosts, vectors, catalysts, and controls. They are elemental aspects of all viral systems.

Hosts, Vectors, Catalysts, and Controls

Whereas cholera sailed and steamed around the world, cholera's children fly.

—Sonia Shah (2016, p. 50)

...the way that anything contagious spreads through a population—be it a virus or a meme—depends on the structure of human social networks.

—Laura Spinney (2017, p. 279)

As an anarchist-feminist organizer (I've heard the jokes), I've worked hard to catalyze campaigns and actions that made direct democracy, personal empowerment and feminist practices go viral. I started with protests for free speech in high school and in college against the Vietnam War, for farm-worker rights and opposing Apartheid, nuclear weapons and power, and other injustices and assaults on our living planet. I've been beaten and/or arrested several dozen times, but the really hard part of social change is growing a resistance community. Even before I knew the terms, I learned it involves these four concepts. Radical ideas and emotions spread by vectors through growth mediums (hosts) facilitated by catalysts or impeded by anti-virals (controls). The specifics vary by the type of infection. Using these four principles, one can accelerate or resist viral proliferation.

Hosts

All viruses need hosts, that is where they live. The most common host for SARS-CoV-2 is *homo sapiens*. After all, there are over 8 billion of us. There are more humans than any other large creatures by a long way. The most common other megafauna are our creations: domestic animals and pets, many of whom can also get COVID-19. Among the pets, domestic, and wild animals who host COVID-19 are dogs (Buddy the German Shepherd died from COVID-19 in 2020), cats, rabbits, hamsters, ferrets, minks (600,000 killed to stop COVID spread), pigs, otters, white-tailed deer, hippos, hyenas, tigers, lions, pumas, and snow leopards. But the most vulnerable are us and our fellow primates, such as gorillas (Rozsa

2021). Only humans "get" cultural viruses. Only the machines humans make get digital viruses.

Vectors

Think of vectors as lines going in one direction through time. Time's arrow is our vector to the future. Vectors can transmit viruses or genetic changes, carry digital information, or guide ideas into people's minds. There is a growing realization about their importance based on their central role in all types of viral spread.

Mckenzie Wark (2019) has even argued that the vectors information travels over in the digital world are now the dominant economic form and so Capitalism is dead, supplanted by a new order dominated by the vectorist class who control the information sector and finance. While this theory hasn't really caught on, it does illuminate a sea change in the dynamics of exploitation, which others have called platform capitalism, cognitive capitalism, and informational capitalism. My preference is for surveillance capitalism (see Chapters 6 and 7).

The philosopher Sean Cubitt (2011) credits Paul Virilio for some of the growing interest in vectors. Cubitt shows their relevance to a wide range of aesthetic and political issues, and well as their central role in biology, mathematics, and media theory. The vectors of SARS-CoV-2 are human transportation systems, gathering spaces (rallies, restaurants, protests, clubs, and contact networks) and person-to-person interactions that produce infections. The host medium is bodies, mainly humans but also bats, pangolins, and a small number of cats and dogs. Catalysts accelerate the spread of viruses, and they are often human predispositions, habits, emotions, and actions.

Social vectors drive biological infections to the viral underclass, through racism, "the liberal carceral state," ableism, capitalism, borders, austerity, and other factors which Steven W. Thrasher details in his viral class analysis of "the human toll when inequality and disease collide" (2022, p. 13).

According to the social epidemiologist David Centola, the spread of behaviors is governed by four factors: strategic complementarity, credibility, legitimacy, and emotional contagion. Strategic complementarity is called the fax effect in technology diffusion studies. The more people who had faxes the more valuable it was to have a fax, until they were rendered obsolete, of course. Credibility is the same dynamic in terms of people's beliefs. The more people who believe something, even dangerous nonsense, the more useful it is to believe it yourself...at least in the short term. Legitimacy comes from more and more people adopting a behavior or emotion, excitement breeds more excitement, just as spreading beliefs generate credibility and increasing adoption lends credibility (Centola 2018, pp. 38–39).

Centola draws a useful distinction between simple and complex contagions. Simple contagions ("pathogenic, informational and even behavioral") are those that "propagate easily across long ties" as opposed to those that "require multiple sources of reinforcement in order to spread." On Twitter hashtags about movies, sport or style were easier to spread than political labels that could well have social costs (2018, pp. 87–89). On X, right-wing views are boosted from the very top. Central credits Hannah Arendt for noting that totalitarian regimes don't fear weak-tie associations, which can even be useful. But strong-tie networks, where people form "trusted, close-knit groups" are an ongoing threat (Centola 2018, p. 92). Having a network of informants at work and in neighborhoods is a great way to undermine social solidarity. Turbocharge it with a culture that encourages informing on your loved ones and your regime is stronger than ever...until it inevitably collapses.

Empathy can spread complex contagions even through weak ties. It comes down to reinforcement and relevance. The reinforcement is found through a few strong links or many weak ties. The relevance comes from empathy and other emotions. Both are needed to overcome social inertia (Centola 2018, pp. 174–175).

Biological viruses as vectors play a major role in genetic engineering, allowing us to rewrite DNA. As symbionts they play a key role in the carbon cycle of the oceans and the life cycle of parasitic wasps. They are also a driver of evolutionary change. Viruses also help spread genes between bacteria and vertebrates, for example (Villarreal 2008).

With COVID-19, fear, skepticism, and politicized idiocy led to denial and the failure to follow important technological practices, such as mask wearing, that could have slowed its spread. Computer viruses are hosted in digital systems and are vectored through emails and social media because of human choices (clicking on that unknown link; "1,2,3" as a password). Viral ideas only really live in human consciousnesses. They infect them through human interactions empowered by networks of various types: family, friendship, television, and the social/digital. Clearly, viruses are crucial in that natural realm that pretends it isn't—culture.

Catalysts
Some forces accelerate the spread of viruses. The adoption of neo-capitalism and the farm factory method in China, for example, has directly led to more flu viruses. Mike Davis pointed out,

> Human-induced environmental shocks—overseas tourism, wetland destruction, a corporate "Livestock Revolution," and Third World urbanization with the attendant growth of mega slums—are responsible for turning influenza's extraordinary Darwinian mutability into one of the most dangerous biological forces on our besieged planet. (2020, p. 51)

When most poultry was produced on small farms, scientists labeled bird (avian) flu an "infection of minor importance." But by 2009 almost 70% of China's eating chickens (broilers) were produced on farms with more than 2,000 birds. Factory farms with more than a million birds have become common. The reproduction rate soars to ten in areas of intense poultry production (Shah 2016, pp. 88–89). Large scale pig breeding is the main way pork is made today, but it has been around a hundred years. It seems the 1918 influenza virus was produced by the mixing of avian flu from migrating birds with porcine flu in pigs farmed in the Midwest of North America.

Urbanization has long produced the same effect. Increasing the concentration of their host growth medium (people, rats, computers, social media platforms) is a catalyst for the spread of all viruses, whether they are biological, digital, or cultural. Climate change is another key factor. As the journalist Chris D'Angelo notes, "Climate Change Is Supercharging Most Infectious Diseases." Research by a team of scientists at the University of Hawaii at Manoa found that our current climate crisis (with its increased fires, changing rainfall patterns in the form of superstorms and droughts and so on) is exacerbating the spread of over half of all important human infectious diseases, from West Nile to HIV. The lead author, Dr. Camilo Mora, admitted "as this database started to grow, I started to get scared, man." Then it became crystal clear that greenhouse gas emissions can impact 58% of important epidemic diseases. When he realized the "magnitude of the vulnerability that we are under" he "went from excited to terrified" (D'Angelo 2022).

Viral ideas are certainly major forces that change culture, as the next chapter shows. But they aren't the only ones. Nonhuman nature—climate, disease, animals, and plants—can certainly change culture. But most other drivers of cultural change can be traced back to culture itself. Technology, for example, which starts with inventors who love knowledge and/or want to cash in, catalyzed through investors with their insatiable hunger for profit, then promulgated by "super spreader" vectors in the form of early adopters and the early adapters who take up, and often modify (mutate), the tech for their advantage. All hosted in the eco-system of the semi-capitalist political economy of today, where competition intertwines uncomfortably with the corruption of politics.

Digital and other new technologies have greatly increased communication between humans, catalyzing new cultural viruses. Douglas Rushkoff was among the first to notice malicious viral media and media viruses. "People are duped into passing a hidden agenda while circulating compelling content," he argued (Thrasher 2022, pp. 54–56). Steven W. Thrasher notes that "...especially when it is *about* viruses, viral news does not transmit in a neutral manner. It mutates and infects along lines of race, class and sexuality making the most defenseless people ever more vulnerable as it reproduces" (p. 55). Thrasher names this "squared virality" (p. 4), and it is a distinct aspect of our age—the digital-to-cultural connection has joined the cultural-biological network, with all three domains driving accelerating feedback to the other two.

Media interacts with emotions and intensifies their impact (Boler and Davis 2018). Emotions are the power driving fake news because it certainly isn't logic or evidence. But not just any emotions, fear seems more powerful than empathy when it comes to reality-free believing. On one level fear can be relatively harmless, if pathetic. Peer pressure, sexual insecurities, smelly body recesses are all weaponized to sell things. Marketing is now using neuroscience to better infect our thinking. Fads often arise out of the chaos of society that aren't linked to sales, they are a feature of our performative sociality.

When it comes to cultural viruses, fear of death is the original and still champion catalyst. It attacks rationality, pushes people together or further apart, denial leads to repression, which returns to supercharge hatred and aggression. Fearful people recommit to their identities, especially those that offer protection from a changing world.

The neurologist Bobby Azarian has a Terror Management Theory he claims explains fantastic MAGA (Make America Great Again, the slogan of the Trump presidential campaigns) beliefs. It is rooted in the *Denial of Death* argument—that humans individually and collectively are terrorized by their inevitable deaths. This was adapted by Sheldon Solomon arguing fear of death makes people more authoritarian: "The more people think about death the more they think about voting for Trump." Solomon predicted COVID-19 deaths of family or friends would propel many Trumpers to double down, not change their minds.

The kind of person who is likely to be a Trump extremist is also likely to have impaired or suboptimal brain function in an important region known as the prefrontal cortex. A healthily-functioning prefrontal cortex is what allows one to override their primitive instincts, to think rationally, and to respond to stressful events in a controlled manner, rather than being controlled by fear and reflexive behavior. (Azarian 2021)

He goes on to point out that "in super stressful times, like during a pandemic, we all become mentally ill in some way," since anxiety and depression certainly increase. We have less control "of our biases and behavior." He advocates counter bias training, psychedelics, and even getting Trump cult members to join his Omega Cosmic Religion for reorientation. Religion is the main way most people manage their terror of death, after all (Azarian 2021).

But fear can become unmanageable. It can paralyze a person or turn into aggression. In the United States, terrified gangs of white men with big guns often confront Black Lives Matter and other unarmed protesters. Their fear is palpable. It is infectious.

Racism is clearly a catalyst for spreading biological viruses. From early in the pandemic it was clear that people of color, a structural part of the viral underclass, are far more vulnerable to contracting and dying from COVID-19. In the UK their death rate was double the rest of the population

(Siddique 2020). U.S. death rates are similar. When there is data. Native Americans were not even a category in the data from half the states in early reporting (Nagle 2020). There are a number of factors in play: poverty, occupational concentrations, access to health care, sexism, and racism.

Controls

Along with hosts, vectors, and catalysts, there are always controls, which can be thought of as anti-virals. Controls block infections from hosts, they cut vectors, they repress or even counteract catalysts. Examples include courage, violence, legal systems, vaccinations, quarantine, isolation, and surveillance. Collective self-control is called community, even democracy. Thought control can be self-control, or it can be external, so-called mind control, or its weaker cousins: manipulation, seduction, selling. If we can manage our fears we can pursue our dreams, we can valorize our better angels. We can decide to follow the advice of scientists and doctors. It doesn't have to be others controlling us—demons or angels, demagogues, or heroes.

Control can come from infectious fear. There is good evidence from species as different from us as the sociable Caribbean spiny lobster who won't den with those infected by a virus and badgers, who hide in their tunnels when exposed to diseased zoo mates. Disgust has evolved in many creatures to limit the spread of infections (Spinney 2017, p. 90). This is no different than how we react when a housemate has COVID symptoms or a stranger is throwing up in the street.

Control can actually be achieved through accommodation. One reason bats are such effective hosts for viruses is that instead of unleashing immune system warriors (white cells, inflammasones, T-cells) to attack invading viruses, they often don't respond when they detect them. They've evolved a way to extinguish the free radicals crucial for accelerating infections. The effect of this is longer life but also accepting long-term slow-burn viral infections.

Since bats are common and varied they often spread diseases to humans and others. Being creatures of the night, looking like flying rats, chomping insects, and occasionally biting humans, all while making strange noises, they are widely feared and hated. Bat persecution was an issue during the COVID-19 pandemic. The fear spreading with the virus led to bat massacres in Peru, India, Australia, Indonesia, Rwanda, and other countries. Even though bats are important pollinators, the agave and baobab trees are only serviced by them, for example, so wiping bats out would be a major disaster. But fear has its own logic (Mackenzie 2020, pp. 122–125).

Viruses are not good or evil. A virus can be a control or a catalyst or a vector for another virus. Most don't affect us, and many are crucial to humans and the biosphere. Often, viruses are best for controlling viruses, as with vaccines. We judge viruses from our human perspective—good, bad, or indifferent. This is particularly true of viral ideas.

3

Viral Ideas

Conspiracies: Burning Books, Bodies, Minds, Forests

The books were burning badly.

—Manual Rivas (1988, p. 32)

It was plain to see that spring had spent itself, lavished its ardor on the myriads of flowers that were bursting everywhere into bloom, and now was being crushed out by the twofold onslaught of heat and plague. For our fellow citizens that summery sky, and the streets thick in dust, gray as their present lives, had the same ominous imports as the hundred deaths now weighing daily on the town. That incessant sunlight and those bright hours associated with siesta or with holidays no longer invited as in the past, to frolics and flirting on the beaches. Now they rang hollow in the silence of the closed town, they had lost the golden spell of happier summers. Plague had killed all colors, vetoed pleasure.

—Dr. Rieux in The Plague *(Camus 1975, pp. 112–113)*

In Santa Cruz it is almost always spring. Winter is some extra rain and summer is extra dry, but something is always blooming and most trees are ever-green. But the least spring-like season is during a drought in the late summer or early fall. Then we burn.

On August 16, 2020, a thunderstorm generated almost 11,000 bolts of lightning, but little rain, across California starting several small fires in the Santa Cruz mountains. Driven by strong dry winds, they combined a few days later into the CZU Lightning Complex fires. Eventually, over 85,000 acres burned and 1,500 buildings were destroyed. Along with the dozens of other fires in California at the same time (LNU, SCU, and SQF Lightning complex fires), the CZU blaze changed the very air we breathed…and much else. As I wrote my ex after the worst had past:

Most of Ben Lomond is safe. Our old street should be fine...Santa Cruz seems pretty safe. They just got a fire line in north of UCSC between 1 and 9 and are putting in a back up.

Bonny Doon, on the other hand, got really fucked. A friend of mine had his house saved by a neighbor while he was saving 3 others. And Swanton Road, and Hungry Gulch (is that what it was called?) got nailed. But it is hit and miss, so maybe our friends have their houses. Maybe not.

Western Drive is on high alert as it is right on the edge of town. But now with marine air and more fire fighters they should be ok. Unless fires start right next to them and aren't caught. Could happen. I'm sure folks living there will fight to save their houses if it came to that. While the fire fighters want everyone out, when they are so overwhelmed as they were the first few days I think that was a mistake. Lots of houses in the mountains were saved by neighborhood teams. And people are reporting that some fire crews let them do backup and mop-up type work.

Lots of ash in town. Bad air. I'm going over to SJ but will be back tomorrow when another lightning dry storm might hit (20% chance). That could be bad.

Santa Cruz, always keeps one on their toes.

I'm not thinking we'll have to evacuate or anything. Don't know what I'd save, actually. I am taking my passport and digital files over to SJ, just in case.

Santa Cruz survived, but more fires are inevitable. In January 2025 it was two whole neighborhoods in Los Angeles that burned. The flames will return to Santa Cruz some day, we know. These giant fires are a plague spawned by the warming of the world by the carbon-economy and spreading where there is wind and fuel. Months after the CZU Lightning fires were defeated, burning logs were found deep in the Redwood gullies that are a major feature of the local mountains. Reinfection/re-ignition is common.

Montana, where I lived for nine years, also knows about fire. Fires follow the same rules but burn differently in Montana than California. The fuel has been shaped by climate and culture. Montana's most famous fire is Mann Gulch, where most of an elite team of smokejumpers burned to death. They were killed by a small grass fire that went exponential. Only 3 of 15 survived. Two had bolted quickly when the fire jumped them and outran it to a ridge. The other survivor set a back-fire that burned a safe space around him. Thus, the escape fire was invented. Dr. Don Berwick, President Obama's director of Medicare, used Mann Gulch as a case study on dealing with epidemics. His four most important principles were: (1) Don't wait for the smoke to clear; once you can see it is often too late; (2) by the time you run from an epidemic it is too late, it is already there;

(3) only do the important; and (4) invent an escape fire, if you can (Lewis 2021, pp. 170–173).

Cultures, and the viruses they harbor, follow the same rules but with variations. The ways cultures are inflamed is more complicated than how forests or grasslands burn or even how cells are turned into inflamed virus factories.

While SARS-CoV-2 cannot survive long on surfaces, some ideas frozen in stories can reinfect after hundreds or thousands of years. Ideas are part of the viral world as tiny mutating parasites that spread and fade exponentially. But there is only so much space for so many ideas in human consciousnesses, even if there are 8 billion of them. Some viral ideas run mindlessly on digital systems, but many others sit dormant in books or scribed on monuments. Many ideas have "died" altogether, as local knowledges, thousands of languages, and billions of people have passed.

Most knowledge isn't viral but even ideas that aren't coexist with viral spreaders. All ideas are part of the same ecosystem and impact virality, being repulsing or inhibiting (controls), accelerating (catalysts), spreading in networks (vectors), and growth mediums (hosts).

Looking at one of the more viral forms of knowledge, conspiracies, is a good way to understand cognitive infection. Conspiracy, from the Latin *con spire* ("breath together"). In common usage conspiracy theories are always crazy. But in the real world this just isn't true. Aren't most human projects conspiracies? Black Lives Matter is a conspiracy. Fascism still is a conspiracy. Conspiracy is what humans do. Breathe together. Demand justice, together. Burn books, together. Raise barns, together. Tell lies, together. Do science, together. Commit genocide together.

The hard part is choosing the right conspiracies to join, which means choosing the ideas we will propagate and act on. It is a choice, even if we often default to the simplistic viruses that match our prejudices, mirror the beliefs of our friends, and don't generate unbearable cognitive dissonance. All cooperation is a conspiracy and we could not survive without it. Then again, conspiracy theories about reptilian baby eaters being battled by secret armies led by dead Kennedys (not the band) and easily traced to anti-semitic lies over a thousand years old, that make people refuse vaccines, consume horse medicine, and support semi-fascist wannabe dictators who run cons but don't govern, are not helpful.

Much of Chapter 5 focuses on QAnon and related mad ideas, but first it is important to understand some general principles of conspiracy theories. Groups don't always exchange information for its truth value. It could be that the information is nonsense but the sharing of it is good for mobilizations in conflict with other groups, for coordinating attention and signaling commitment. Labeling your enemies pure evil with no evidence (pedophiles!), believing something most people won't (The Big Lie!), and performative dress and actions to offend outsiders ("to own the libs") fosters group cohesion. This is one reason our reason can be so unreasonable (Rosenberg 2021).

Why are some implausible, impossible really, conspiracies taken up and not others? Why do some people believe in truly insane conspiracies and other don't, and why do some believers act on them and most do not? A large study of Reddit users in 2019 showed that "stressful life events" could overlap with existing feelings of powerlessness and distrust to make people more vulnerable to irrational claims. We also know that believing in one conspiracy theory, reasonable or not, increases someone's chance of following others. Being in a group, homophily ("birds of a feather…"), can be particularly distorting (Klein, Cotton, and Dunn 2019).

It is also useful to differentiate conspiracy theories from conspiracy practices, as Edward Snowden—no newbie when it comes to conspiracies—argues. He is struck by how "the truest conspiracies meet with the least opposition." These are the giant collaborations that "order, and disorder, our lives" while enriching the 1%. Snowden realized that many people wanted to talk about conspiracy theories because actual conspiracy practices were "too daunting, too threatening, too total." For him, conspiracy theories are a symptom of powerlessness. After all, he started working for one well-known "top secret" conspiracy (the U.S. national security state) and joined another much smaller one (opposition to the same). It remains to be seen if getting Russian citizenship has reenrolled him in another (Snowden 2021).

Salon columnist Chauncy Devega (2023) makes another useful distinction, between conspiracy and conspiracism. Conspiracy is "two or more actors working…to advance their own interests." Conspiracism is a framework, a "theory of knowledge" so that "one's understanding of events is understood" only through that conspiracy. It "is a meta-conspiracy theory." He quotes other analysts on the "asymmetrical conspiracism" that has taken over the Republican Party, fueled by "the fantasy industrial complex" of social media and the grifters who use it to monetize their bizarre conspiracies.

So it should be no surprise that faced with a deadly pandemic on top of a wildly careening political crisis with the dark clouds of climate change and weapons of mass destruction overhead, while the rich get richer and everyone else poorer, many people turn to conspiracy theories for solace and solidarity. It doesn't mean they are not true. We need to analyze them. A good start is not attacking conspiracy theories immediately but asking how can we tell which ones are probable, possible, or pathetic?

First, consider if it is true how many people are in on it? Trump's Big Lie about his 2020 election loss requires hundreds of thousands of conspirators—not just Democrats and independents but many Republican judges, governors, and election officials, and, of course, most intellectuals and media workers. Not very likely. Most of the craziest pandemic conspiracy theories (secret conspiracies, not Big Pharma, for example, a very public collaboration to make money at the expense of people's health and lives), postulate the cooperation of hundreds of thousands of scientists, doctors, and other medical and public health workers, elected officials, the courts, and all the "lame-stream media." Pretty much impossible. Now, just because a conspiracy is small it doesn't mean it is real, but the paranoid ravings of QAnon believers and their ilk are clearly not credible.

Next ask, *cui bono*? Who benefits? You can't always trace the author of some disaster to who benefits most, but you often can. If the supposed conspirators get no benefit from it, you have to wonder if they really did it. Why would the vast majority of medical professionals in the world foist vaccines with mind control chips in them on the rest of us? What do they get out of it?

Another good question about a conspiracy is does it show resilience when parts are debunked or does it depend on many isolated and improbable connections that need to all be true for it to be real? For the Oswald-as-Communist-Gunman theory of President Kennedy's assassination to be true you have to accept that he was an amazing sniper, that "the Magic Bullet" hit both Gov. Connally and Kennedy making a 90-degree turn in the process, that Oswald's defection to the USSR was entirely normal even though he was a Marine with a high security clearance who wasn't even arrested on returning to the U.S., that he was killed by the mobster Jack Ruby out of an excess of patriotism, and that the mishandling of President Kennedy's body and losing his brain was just bad luck. If any one of these very shaky claims is disproved, the Oswald theory collapses.

This is all well and good if you are trying to analyze the possible validity of a conspiracy theory you come across. But some people aren't looking for logical satisfaction when they embrace conspiracy theories, they are looking for emotional validation of their hatreds and fears.

Where do these theories lead us? To burnings…bodily fevers, inflammations of the body politic, books and people incinerated at *Auto de Fé's* (act of faiths). The Earth overheating is in large part because of the very real conspiracy of coal and oil companies maximizing their profits while spreading fearful fantasies of conspiracies of scientists taking over the world by faking global warming. Book burnings are a symptom of the disease of authoritarianism. Body burnings are another. The ultimate aim is mind burning, as one used to "burn" a disc of data. Just as many biological viruses produce burning fevers, cultural viruses can produce the overheated thinking that leads too actual burnings of books, minds, and bodies. When they occur together they feed on each other.

In the case of COVID-19, one snippet of reified fear argued that 5G towers were the actual source of the infection, so across the world they were put to the torch, with dozens attacked in Europe, over 60 in the UK. There were attacks in the U.S. as well, including Tennessee, West Virginia, and New York. Most notably on Christmas Day 2020 one believer, Anthony Quinn Warner, destroyed the AT&T building in Nashville and himself with a van bomb (Klippenstein 2021).

There wasn't just one theory among attackers. Some claimed the virus itself somehow came from the towers, others that the 5G rays weakened us for the virus, still others that the towers were killing us and there is no SARS-CoV-2 at all. How do they know? Well, on the new £20 note there is a 5G tower! Or is it the lighthouse in Margate, so beloved by the painter M.W. Turner, whose face graces the bill? It doesn't matter really. Fear trumps evidence (Subramanian 2020). This kind of nutty thinking gives conspiracy theories a bad name and sustains authoritarianism.

In his novel about Galicia focused on the fascist book burning on La Coruña's docks in 1937, Manuel Rivas tracks the lives of a number of people involved in resisting or perpetuating this famous act of faith. He shows how burning heretics and their books is about spreading certain viral ideas, not just eradicating dissident viewpoints but transforming feverish fear of the other into love for the incendiary authority.

I've mixed feelings about fevers. They can be cleansing. That is what they are for, overheating the cozy environment where bacteria thrive. But if the fever threatens my sanity (Istanbul 1972), or comes with a great deal of other unpleasantness (Mexico and Egypt several times over the years and norovirus in Santa Cruz in 2019), what can you do? You hang on.

When the COVID pandemic hit, Bathsheba Demuth was in Siberia studying the historical impact of Yankee whalers in the 1800s, so she hung on. She got long COVID-19 in the early summer of 2020 in a land that was suffering its own fever of massive fires—of forests, grasslands, and even tundra—that burned over 50 million acres and produced as much carbon dioxide as Portugal or Sweden do in a year.

My body has been alight for months now. From within this illness, I have come to think that Siberia and I endure more than a coincidence in temperature. Our fevers are stoked by related patterns of economic production, patterns both relatively new and seemingly inevitable. And my corporeal fire says something about how a continental fire can go unseen, offering a lesson in the implications of duration: how as a condition lingers, its origins or significance grow harder to see. Long COVID and climate change are alike in this: live ill for long enough, and the absence of health threatens to become normal. (Delmuth 2020)

But despite her own illness, and the near apocalyptic conditions she found herself in, she finds some hope.

My experience of this virus makes me think, however, that we should not forget a longer view, one able to see how the conditions of 2020 are not inevitable. The line of heat that connects my body and Siberia has existed for only a few centuries. It is not inevitable. Thinking past it, as this summer of our many discontents moves into fall, requires a kind of split imagination: to conjure moments of past flourishing, and a future where we might flourish again. (Delmuth 2020)

We don't have to be the victims of the conspiracies of greed that are burning up the world; we can conspire ourselves to flourish. But carefully. We need to remember that maybe trying to do good led to the pandemic.

Science and COVID-19

Today, the notion of virality animates the research agendas of hundreds of thousands of scientist worldwide, ranging from computer scientist and physicists, to sociologists and marketing scholars.

—Damon Centola (2018, p. 2)

Since the mid-twentieth century, biomedicine has been rightly celebrated for its powerful ability to render lifesaving cures. But its limits, which have already started to show, will only become more apparent in the coming years. Some of the external disruptions that are now eclipsing microscopic mechanics as drivers of new disease are more amorphous, wide-ranging, and unpredictable than the world has ever seen.

—Sonia Shah (2016, pp. 162–163)

In academia, my field is the Cultural Studies of Science and Technology. I also study science as my avocation. I love science, I understand science. Science is a very powerful but it isn't perfect, in fact it is always wrong. But its process is about getting less wrong with each experiment. One needs to keep an open mind. Being sure that Western "scientific" medicine is an evil conspiracy or that it has all the answers are both ways of getting dead before you need to. After all, the lines between today's science and other ways of understanding the world are far from clear.

Many "nonscientific" medical theories and practices actually come from the traditions of indigenous cultures throughout the world. The "germ" theory of disease took centuries to replace the ancient view that dis-ease was a matter of imbalance among basic forces whose harmony was health. Humors, in the framing of the West's Hippocratic medicine (*miasmatism*), the principles of *vata*, *pitta*, and *kapha* in Ayurvedic systems, or yin/yang (China) could be unbalanced by bad air, miasma, or other forces and rebalanced through herbs and food, physical interventions such as massage, acupuncture, surgery, or meditation. The transition between this holistic medicine and the germs-against-drugs-and-machines medicine we have now was not easy.

Despite the evidence from Louis Pasteur that diseases could be caused by bacteria, and Filippo Pacini and Robert Koch discovering *Vibrio cholerae* and showing it caused cholera, most medical experts in the 19th century did not change their view that germs either didn't exist or were not important. One scientist, the chemist Max von Pettenkofer, went so far as to drink cholera to prove germs harmless, joined by his assistant. While they both became violently ill, it only lasted a day or two so he considered it proof of the irrelevance of microbes. But after decades of arguments an outbreak in Hamburg that spared its suburb Altona, which filtered its drinking water, convinced most skeptics (Shah 2016, pp. 156–157).

Sadly, when a major paradigm shifts in science often the old system's insights are not retained. In *Pandemic*, Sonia Shah explains that "Revolutionary new paradigms generally don't accommodate old ones, subsuming their principles and approaches. They destroy the old ideas and purge the ranks of loyalists." For her this is personal. She learned to take a more balanced view during her family's experience with the superbug methicillin-resistant *Staphylococcus aureus* (MRSA). The first infected was her 13-year-old son with "angry" boils on his knee. After a wide range of antibiotics and painful draining treatments his leg was saved. But, the doctors warned, it might reinfect him or someone else in the family (Shah 2016, pp. 9–10).

It did. A few months after his first bout the infection emerged again, and he had to do another round of "semi-toxic antibiotics." Later another infection irrupted. It, too, was defeated. Then Sonia Shah was infected herself (Shah 2016, pp. 33–34).

Her struggle with MRSA lasted over three years. She tried everything her doctors recommended, but she noticed that they "didn't ponder the landscape, or my home environment, or my immune statues, or the animals that lived in my house, or my diet." Instead, they "targeted the bug and only the bug." In their view it was a gunfight between MRSA on "one side of an invisible divide" and her "on the other, gun in hand." She found this reductionistic, exactly "opposite" of Hippocratic medicine's "fundamentally holistic and interdisciplinary approach" (Shah 2016, p. 160).

Finally, after years of aggressively confronting the returning sores with "sterile pads, tape, antibiotic cream, and drawing salve" and "boiling of clothes, more wiping of counters, more drugs, more sprays, more bleach baths to rout out the intruder" she "stopped fighting." She was tired. The next time the infection appeared, a little bump, she did nothing. She did not "squeeze or apply ointment or heat or bleach." She didn't scratch it. She ignored it. It went away. She was cured.

She admits, "I have no idea why this happened" but does "suspect it had to do with more than just the microbe my doctors and I had surgically focused our ire upon." Perhaps, she muses, "There was some kind of Hippocratic interplay going on, between internal factors and possibly external ones, too. As she notes, "Modern medicine, singularly focused on the microscopic, is poorly suited to grasp such interactions" and yet that is what we are faced with in the real world, especially now in this Age of Pandemics (Shah 2016, p. 161).

Shah cites a recent survey that showed biomedical experts "rarely collaborate with social scientists." They actually don't play that well with each other. As powerful as biomedicine has proven to be, it alone isn't enough to deal with the diseases we face now in our polluted, overpopulated and overheating climate, "new disease are more amorphous, wide-ranging, and unpredictable than the world has ever seen" (Shah 2016, pp. 162–163).

Climate change has driven bats, mosquitoes and ticks to new environments which haven't faced the pathogens they carry before. Yellow Fever has appeared in California, Lyme Disease all over North America, West Nile across the continent as well, and it is the same in other places. Fungi, which can be "potent pathogens" are also an emerging threat, especially if they evolve to stand heat of 98.6 or so degrees, as they seem to be in our ever warming Earth. As the microbiologist Arturo Casadevall warns, "Heat-tolerant fungi, if they emerge, would be an infectious disease threat like no other" (Shah 2016, pp. 174–176).

SARS-CoV-2 came from harvesting wild animals, after all. Or not. Inevitably there are many conspiracy theories around the origins of COVID-19. Most involve confusing the occasion with the cause. Even if it escaped from a lab doing animal passage/gain-of-function research, that was not the cause of the pandemic, merely the occasion. Some kind of pandemic had been overdetermined for some time because of the proliferation of people, the exploitation of the wild, capitalism and other extractive profit systems, and scientific practices. Civilization produced the COVID-19 pandemic. More pandemics are inevitable.

But did it come directly out of our relentless consumption of wild nature or from scientific work on viruses? There is a great deal of evidence for the theory that the virus passed to humans through the Wild Meat

Market in Wuhan. But it is also possible SARS-CoV-2 came from the lab. We may eventually know, new genetic or political evidence can be expected, but it doesn't matter.

Whether or not this epidemic came out of Wuhan's labs, it is clear now that it could have. A standard, but controversial, approach to understanding viruses for treating the diseases they cause, for weaponizing their bad health impacts, for creating vaccines or using the viruses as vectors for delivering something biologically, is called gain-of-function research. Scientists actually try and enhance transmissibility from one species to another and the pathogenicity of the virus in order to understand it. This work happens in Wuhan's lab, funded in part by WHO money, as it goes on in dozens of labs around the world (Guterl, Jamali, and O'Connor 2020). If it was a lab accident it could just as easily have been a U.S. lab. All labs make mistakes, eventually.

Gain-of-function work is controversial. Some scientists have argued that it is too dangerous to do (Duprex et al. 2015). Supporters respond that it is the best way to develop the knowledge needed to fight pandemics such as COVID-19, including producing vaccines. A full gain-of-function experiments debate, as Paul Duprex, Ron Fouchier, and others urged in 2014, is necessary. It is dual use research of concern (DURC), which WHO defines as "life sciences research that is intended for benefit, but which might easily be misapplied to do harm" (WHO 2020). The COVID-19 pandemic has led to renewed attention on DURC. Are the benefits of gain-of-function experiments, such as improved vaccine production, outweighed by the danger of generating pandemics or making bioweapons? What if searching for cures spreads diseases?

Spreading Vaccines

Should we recoil in horror...from the fact that our bodies are programmed with simulated diseases and that the simulated can sometimes kill? The programming of the human body has just begun and programming errors are inevitable. The recoding of our immune systems by vaccines is just one of the ways we have made ourselves cyborgs. Should we recoil from ourselves and our children? Of course not, a parent answers, but it is something that should not be ignored.

—Chris Hables Gray as a graduate student and new parent (1985, pp. 32–33)

Preventing a flu epidemic that could kill thousands is not nearly as profitable as making pills for something like erectile dysfunction.

—Donald Barlett and James Steele (Davis 2020, p. 154)

One of the most immoral things I've ever done is not getting my sons vaccinated when they were little. This wasn't because I didn't believe in vaccines but because I did. For the mandatory baby vaccines required by California, my research showed that herd immunity was a safer choice than the actual shots which went to every child except those with religious exemptions. Vaccines do have real dangers after all. Just read the labels. Yet while some people will get seriously ill or worse from vaccines, usually many more will benefit from not getting the disease at all, or at least a less severe case. So, I proclaimed, as pagans, it was against our religion to get vaccinated—modifying the body the Goddess gave us. OK, I lied. At the time, the U.S. used a combo measles/mumps/Rubella vaccine that was banned in most of Scandinavia for being too dangerous. It was later discontinued. Eventually, my boys got the vaccines they wanted when they were older, as vaccines improved and risks and travel arose.

Why was it so immoral? Because I put my family above society, a path that if followed by more people leads to a failure of vaccines to protect society. I admit it, having kids threw me a bit. I've tried to balance a commitment to helping humanity survive with a good life, but with my sons I've been prejudiced. In my romantic choices I'm often bat shit crazy but never at the cost of my politics, my relation to the *polis*, to my fellow humans, to community. But as a father I found it harder to be moral and rational.

I was first vaccinated before my family moved to Vietnam. I was four. My two brothers and I got lots of shots in our butts. The second shot session at the doctor's office I escaped and hid under an exam table. They caught me in the end, pun intended. It is one of my first memories. More happily, I also remember getting the first dose of oral polio vaccine soon after we moved to San Diego. Such a relief! I remained terrified of needles into my twenties.

Vaccines have come a long way since people were given cow pox on purpose. The basic principle is a vaccine "teaches" the immune system how to fight the targeted disease. Early on it had to be live but weakened versions or relatives of the actual disease, like cowpox is similar to smallpox.

Even now smallpox vaccinations protect against monkeypox. Technically, only some vaccines are viruses—those created with attenuated (weakened) viruses such as chicken pox, mumps, measles, and rubella. But in terms of their complexity and how they are spread, the viral aspects of all vaccines are clear.

Scientists have developed many different ways of using parts of infectious agents with enough similarities to the targets to stimulate effective immune responses. "Killed" (inactivated) vaccines, such as whooping cough, use triggering parts from a bacteria or viruses. Toxoid vaccines use a toxin or chemical made by the virus or bacteria. Tetanus vaccine is the most common of this type. Now vaccines can use snippets of DNA or even messenger RNA to provoke the immune system.

Understanding probability is crucial for speaking Virus. There is no 100% guarantee a vaccine will keep you from catching a disease, or keep you from getting sick if you do catch it, or even save you for sure from dying. But effective vaccines greatly increase your odds.

Some contagions are more complex and need multiple exposures to infect, but they aren't fundamentally different than simple ones. Multiple exposures always increase infectivity, just as someone becomes more likely to quit smoking as more of their friends do. It is the same for catching most biological viruses.

Virulence and immune response are crucial. The emotional intensity of the exposure is particularly relevant for cultural spread, but similar dynamics apply to machine systems. Digital infection vectors are often emotional ("YOUR COMPUTER HAS BEEN INFECTED! Click here!") as well as algorithmic. The more infections the greater the chance of more infections, in social, biological, and digital systems. Until, that is, rising levels of immunity or counter measures apply.

Herd immunity is hard, often impossible, to achieve. Vaccine shortages, vaccine resistance, and a virus that mutates easily, fostered the COVID-19 Pandemic. Even a 70% vaccination rate would not have been enough for herd immunity. Vaccines can't prevent 100% of transmissions because they aren't perfect and infectious beasties are constantly mutating. Viruses thrive in the gaps. New variants shift realities. Delta's heightened ability to reproduce and spread shifted the herd immunity rates drastically. Acceptance of vaccines is very uneven. Some places had very high vaccination rates and others quite low, either because of vaccine fears or because whole countries had trouble getting enough vaccines (Aschwanden 2021). Hoarding vaccines by rich countries killed 1.3 million people, according to a study by University of Warwick epidemiologists (Farah 2022). Add them to the many millions in the viral underclass who die unnecessarily every year.

Unsurprisingly, COVID-19 made vaccines a major political issue. There have often been paranoid conspiracies about vaccines. In 1998 when WHO was trying to eradicate polio, Muslim leaders in Nigeria said the polio vaccine "was contaminated with HIV and secretly meant to sterilize Muslims." In North Waziristan, Pakistan the campaign was halted for a year. Over in Afghanistan, Taliban leaders announced it was cover for espionage. In Muslim parts of India it was claimed that the polio "shot was contaminated with pig's blood and contraceptives." In Nigeria health workers were assaulted and in 2014 in Pakistan 65 health workers giving vaccines were murdered." Polio surged and WHO declared a global health emergency. Eradication had failed (Shah 2016, pp. 134–135).

Vaccines can be power. China used CoronaVac and Sinophar for foreign policy as well as domestically. Almost half of their billions of doses were distributed outside China (Mallapaty 2021). Russia used its vaccine Sputnik as a "hybrid weapon" according to the prime minister of Lithuania, who objected to the conditions Russia wanted to impose for donating doses. Although only 19 million Russians were vaccinated as of April 2021, Russia still tried to leverage the vaccine for political gain (Henley 2021).

When China offered the Philippines their vaccine, the U.S. military started a PSYOP campaign using accounts such as #Chinaangvirus (Tagalog for "China is the virus") claiming China started the virus, and then sold PPE and vaccines to profit from it. It used their "sprawling ecosystem of social media influencers, front groups, and covertly placed digital advertisements." The operation included South East Asia Central Asia, and the Middle East, where the lie that pork gelatin was in the China vaccines was spread. Started under Trump and discontinued by the Biden Administration, the PSYOP probably cost the lives of thousands of people, especially in the Philippines where both the campaign and vaccine resistance was quite strong. It is well established that when trust in one vaccine is compromised, whether merited or not, general distrust of all vaccines spreads. Despite ending this particular campaign, Biden did give the contractor who carried it out, General Dynamics IT, a new contract for similar PSYOPs worth almost half a billion dollars (Bin and Schectman 2024).

As of June, 2022, vaccines had saved at least 20 million people worldwide, according to a mathematical analysis published in Lancet (Watson et al., 2022). With the continued development of targeted boosters, and vaccines that might soon protect not just from SARS' variants but the flu, HIV, and Malaria, debate over vaccines seems destined to continue. Some call it the Golden Age of Vaccines, but others see them as a trick for subjugating humanity.

When you are infected by a new virus you are changed, you become a new kind of symbiont (for you are already teeming with other viruses). But if you are vaccinated you become a cyborg, or if already vaccinated, a new variant of cyborg.

Cultural viruses are just as transformative, just as weird, just as dangerous.

Dangerous Cultural Contagions

It would not be too farfetched to say that the extermination of mankind begins with the extermination of germs. Man, with his humours, his passions, his laughter, his genitalia, his secretions, is really nothing more than a filthy little germ disturbing the universe of transparency. Once everything will have been cleansed, once an end will have been put to all viral processes and to all social and bacillary contamination, then only the virus of sadness will remain...

—*Jean Baudrillard (1993, p. 61)*

All plagues follow bad leadership.

—*Sophocles,* Oedipus Tyrannos, *476 BCE*

Everett Rodgers was one of the first to try and apply epidemiology to culture. He was a professor of Communications at Stanford when I was an undergrad there. One of my favorite protests in the early 1970s was going into his class with several dozen other Stanford students and Iranian activists carrying large signs in English and Farsi denouncing the Shah of Iran. We didn't have to say a thing to "disrupt" the course, because it was being filmed for use in Iran and the messages made this impossible. Rodgers and

Stanford had a contract to "sell" the Shah to the Iranian people through a new network of satellites and sophisticated programming with infectious pro-Shah ideation.

Rodgers' 1962 book *Diffusion of Innovation* argued that—as with the spread of biological infections—successful new ideas and products followed an S (from sigmoid function) curve in their lifecycle: slow start, quick growth, and then a leveling off. Rogers argued that once 20-25% of people had adopted a new product or idea, it had "taken off" and could not be stopped. He also defined the different roles people play in this: innovators, early adopters, majority of the population, laggards. He predicted new technologies (radios, televisions, microwaves, mobile phones) would turbo charge this process (Kucharski 2020, pp. 34-36).

That part of the contract was cancelled but not all of it. Before Stanford could deliver the satellite system and get paid, the Shah was overthrown and Stanford lost millions in stranded costs. Later, during a sit-in to get Stanford to divest from apartheid-supporting stocks, the VP for Finance reminisced with me about the Iran protests. He sighed, "I wish you guys would have won that one." I wish the Shah's dictatorship would have been replaced with something better than theocracy, like the democracy the U.S. overthrew in 1953. We often don't get what we wish for.

In these weird times it is important to evaluate how different political systems perform. With COVID-19 much remains unclear. Some working democracies have done well, others not. Many authoritarians have consciously or with incredible stupidity killed hundreds of thousands of their people and others have clamped down on the virus with all their considerable powers, saving just as many.

Ironically, the success of China's initial response to COVID-19 was grounded in their bio-surveillance regime (Hester 2020), which has been under development for years to control dissidents and conquered ethnic minorities, especially the Uyghurs and Tibetans. It includes facial recognition, fingerprints, and DNA in massive databases that also track legal and credit information, and even the standing ("social credit") of citizens. Perhaps having this capability inspired the government to impose draconian lockdowns for millions into the fall of 2022, making the sudden end of all restrictions in the face of Omicron all the more surprising. While that decision might have cost as many as 2 million excess deaths in China (Jett and Cheng 2023), it still had a much lower per capita death rate than the U.S.

Nicholas Christakis, a physician and sociologist, is scathing in his evaluation of the American response.

Around the country, administrators who should have been finding ways to improve the efficiency of their hospitals or obtain the equipment that physicians and nurses needed to care for patients were instead trying to manage the epidemic by censoring or suppressing bad news. (2020, p. 153)

He links this failure to the disempowering of doctors and other health workers that has gone on for years as profit-maximization has turned them into "mere employees of large organizations that are often led by people who have little appreciation for clinical care." It didn't help that the "muzzling of doctors took place at the highest level of government," and he details the many stupid, even criminal, actions of President Trump. For example, claiming "no-one" could have predicted such an epidemic, promising from the beginning it "would go away" and promoting fake cures and racist conspiracies. The politicized CDC did little better (2020, pp. 152–157).

Gov. Cuomo of New York and the Democratic leadership of New Jersey and Connecticut made mistakes contributing to the worst outbreaks in the U.S., but other liberals did better, especially California's Gov. Newsom. Once vaccines became available all Democratic areas outperformed Republican. There was a clear shift to outbreaks focused around unvaccinated (i.e. Republican) areas. We can measure the actual costs of Republican science denial; hundreds of thousands of Republicans, perhaps a million, died because of Republican policies. Looking at the first three waves of COVID by counties shows heavily urban (Democratic) counties suffered the worst by far in the first wave but once vaccines became available the Republican areas had much higher death rates (Hartman 2021).

Doug Haddix used CDC data to show that the "14 states with the highest death rates were all run by Republican governors." Florida led with 153 deaths per 100,000. Charles Gaba, a health-care analyst reached a similar conclusion looking at the top 16 deadliest states. Florida residents at this point in the pandemic were two and a half time more likely to die of COVID-19 than Californians. The journalist Dana Milbank asked, "How does Ron DeSantis sleep at night?" considering his mismanagement of COVID-19, including lying and the purging of whistleblowers and anyone wanting pandemic policy set by science and not by politics (Milbank 2022).

The Trump Campaign's own pollsters calculated that voters who rejected him because of his pandemic policies were the decisive element in his 2020 defeat. COVID-19 deaths and migration patterns also cost the Republicans the Senate and a number of House seats in 2022. Political analysts call it political Long COVID (Gest 2022).

A task force convened by *The Lancet* argued that 40% of U.S. deaths could be traced to Trump's policies, based on comparisons to other G7 countries. To be fair, they said the rest of the excess deaths in the U.S. could be blamed on the for-profit health system and other social issues (Holpuch 2021). Should the failures of political leaders in the West to respond to the virus be a surprise? Why, as Josef Bouska (2020) asks, is one terrorism death a threat to Western civilization and hundreds of thousands of COVID-19 deaths are just in the way of getting back to normal? In October 2021, the 10% of the most pro-Trump counties had six times the death rate of the least Trumpian 10%, attributable to vaccine misinformation among officials and citizens. In a poll by the Kaiser Family Foundation the same month, 90% of Republicans believed at least one major falsehood about the pandemic (Wood 2021).

Considering Republican leaders were major purveyors of pandemic misinformation this should be no surprise. COVID conspiracies ranged from the vague "Bill Gates owns your body now and he wants you to take the vaccine" (Tucker Carlson quoted in Greenspan 2020) to the scarily specific: "Gates has put microchip trackers in the vaccine," which 44% of Republicans believed in May of 2020 (Greenspan 2021). Pictures of the "'5G Covid' mind-control chip" appeared online, good evidence if they weren't actually diagrams of the Boss Meta Zone Mt-2 pedal chip, a guitar pedal controller (Smith 2021).

Other COVID conspiracy theories were not quite so insane, but they could kill you. At least one person died drinking the fish-tank cleaner chloroquine, and hundreds more might have died from that, or using bleach or alcohol-based cleaners to counter the virus. Ivermectin, horse dewormer, was suggested as a possible treatment by the same scientists who argued for steroid treatments, which did turn out to be helpful. Once ivermectin was tested and found useless, they moved on. But not so for Trump and many of his allies, especially the right-wing Frontline Covid Critical Care Alliance. The more science said it was useless, the more they wanted it, despite the side effects, which are just what you'd expect from a dewormer meant for large mammals (Kabas 2021). Another Trump enthusiasm, hydroxychloroquine, a drug used for treating malaria, rheumatoid arthritis, and lupus, was only effective in killing thousands of people, including more than 12,000 in the U.S. (Plummer 2024).

When virologist David L.V. Bauer gave a video interview about how the Pfizer vaccine worked less well in producing antibodies against the Delta variant than earlier ones, he was shocked to find out that it was soon reedited to make him both a heroic anti-vaxer and a bioweapon-making villain (2021). He is neither. He heads an RNA virus replication lab at the Francis Crick Institute in London and he recommends vaccines. He noted that,

Like the virus itself, the videos seemed to be mutating and spreading, with new, more virulent variants catching on online. One of the most widely viewed videos created a convoluted and conspiratorial narrative involving vaccines, alien DNA and abortion which was repeated over and over – and featured the same clip of me replayed over and over at various points. (Bauer 2021)

Hardly a surprise, as a popular QAnon view is that the vaccine includes "hybrid serpent venom from Satan" (does insurance cover that?) while at least one influencer (Christopher Key) injected himself with "aged urine" because it is obviously better than the fresh stuff, and certainly any vaccine (Einenkel 2022).

As primates, we can learn a lot from studying how behavior spreads among other animals. In the 1940s British researchers noted that great tits widely shared the information on how to peck through the foil on the top of milk bottles to get the cream. Years later, a team of zoologists, led by Lucy Alpin, set up a clever experiment near Oxford where they put mealworms in a puzzle box and studied how long it took for great tits who were taught the secret, a glass door that only slid one way, to get to the yummy wigglers. Despite the high mortality rate of great tits, the secret was passed down in most of the sample groups. The research proved that some great tits were super-spreaders, teaching the method that they learned, even though there were other solutions. There are social norms even among great tits (Kucharski 2020, pp. 94–95).

Adam Kucharski notes that it is often hard to untangle the basic dynamics behind the spread of ideas. It could be homophily, social contagion, or a shared environment…or most likely all three. He cites Max Weber, who noted that when it rains people aren't necessarily opening their umbrellas because people around them are but rather because of the water falling from the sky (Kucharski 2020, p. 95). We can wonder why do they have umbrellas? In Corvallis, Oregon, where I lived for four years, people usually don't carry umbrellas because the rain often comes with heavy wind, making them useless. Besides, they like to pretend it doesn't rain as often as it does. In Great Falls, Montana, where I lived for nine years, it is the same with the added cultural impetus from the orneriness of Montana folks who pride themselves on ignoring the weather. So not only will many people not have umbrellas in a rain storm, they won't be wearing rain coats or any coat at all. Denial is an integral aspect of human nature.

There are even darker forces driving many of the lies about the pandemic. Fear of the virus becomes fear of the other. Russian dictator Putin declared homosexuality an evil virus of love and compared "gender nonconformity and the push for trans rights to 'new strains' of a 'pandemic' much like the coronavirus" (Simon 2021). Chauncy Devega (2021a) calls this kind of thinking "coronavirus fascism" and in the U.S. "gangster capitalism" (neoliberalism) or "competitive authoritarianism." While fascism is a virulent infection indeed, it does not spread easily. As Rivas explains, the infection still lingers in La Coruña, but it is no longer raging.

The book fires are not part of the city's memory. They're happening now. So this burning of books isn't taking place in some distant past or in secret. Nor is it a fictional nightmare thought up by some apocalyptic. It's not a novel. This is why the fire progresses slowly, because it has to overcome resistance, the arsonists' incompetence, the unusualness of burning books. (1988, p. 34)

Although all sorts of horrors can be normalized for at least short periods; repeated often enough, evil can seem to become banal, although it never is for its victims.

Still, biological, cultural, or digital viruses can revive, and go viral. It just takes the right stimulus. James Currier, a venture capitalist, claims there are only eight "psychological buttons" that are behind something going viral digitally. He mined them from 150 million psychology tests from tickle.com. The eight: Status, Identity, Being Helpful, Fear, Order, Novelty, Validation, and Voyeurism. They are all different angles on our existence as social animals. But his analysis is just about "buttons" to push. In reality, these all interact in complex ways. Fear drives us deeper into our identity, and what constitutes high status can shift drastically. Being helpful is a matter of debate—it could be enforcing imposed order or exploring novel assumptions. While people tend towards one or the other, most combine elements of order and novelty. Validation is status and identity, and voyeurism is the drive to observe and catalog others, crucial for social creatures such as ourselves (Stillman 2021).

So what about logic or even self-interest? Are these unimportant for selling or just something sellers would have us not focus on? Because it turns out ideas don't infect that easily. Massive influence campaigns spend millions to impact small percentages to buy or vote a particular way. And one needs a way in. A particular weakness for some is their idea of themselves. When facts challenge their self-image the facts often have to be sacrificed.

Feeling is a type of thinking. Our consciousness is modular. There are many subsystems for feeling and/or knowing different things. But our thinking has to be consistent within reason, especially when we are called upon to act. Cognitive dissonance is the tendency for us to form a unified perspective, even if there are contradictions among facts and feelings.

In "The Role of Cognitive Dissonance in the Pandemic" social psychologists Elliot Aronson and Carol Tavris directly confront one of the major blocks to people changing their minds about it: "The minute we make any decision—'I think COVID-19 is serious; no, I'm sure it is a hoax'—we begin to justify the wisdom of our choice and find reasons to dismiss the alternative" (Aronson and Tavris 2020).

Consider the racist conspiracy theory without any basis of fact that was pushed by a major government insisting the source of the pandemic was a foreign infection. No, not Trump's "China Flu" ravings but the Chinese claim that a white American mother of two, Maatje Benassi, bought SARS-CoV-2 to China, sparking the pandemic China was blamed for! As a reservist and a competitive cyclist, Benassi participated in the Military World Games in Wuhan. She was injured and hospitalized, which apparently was all that was needed for an online troll from the U.S., George Webb, to finger her as the source of China's COVID-19. Chinese posters, including the Communist Party, took up the witch hunt. The Benassi family had to go into hiding due to death threats (O'Sullivan 2020).

Plagues often generate real conspiracies, especially by the authorities trying to keep them secret. When the bubonic plague hit San Francisco in 1900 the mayor and governor denied it. They tried to only count Chinese deaths but were blocked in court. Then they accused the lead federal health official of spreading the plague himself, and one of the main papers, *The Call*, declared the whole epidemic a hoax (Price 2022, pp. 48–50). We are now much further along in understanding how biological viruses work than political infections. Human culture changes incredibly quickly.

In mid-October, 2021 there were still fires in California. A few miles away a small one burned next to my University's campus. Some 20 miles south, a major fire burned outside Watsonville. And 250 miles down the coast, the massive Alviso fire ravaged the old *Rancho Nuestra Señora del Refugio* land grant. Founded by the patriarch of one line of my ancestors, Captain José Francisco de Ortega, Refugio is now a State Park. Refugio: refuge. But there are no refuges anywhere now. Besides, was it a real refuge? There was a pirate attack by the Argentinian Hippolyte de Bouchard, tensions with the dispossessed Chumash, bears, illnesses, high infant and maternal mortality and much shorter lives. And fires. Fires then and fires now. Inevitably, for everyone, death. Death comes to us all. Maybe that's the point. Refuges are only temporary respites. Our ancestors knew this well, but we like to pretend otherwise.

4

Culturing Cultures

Viral Evolution and Civilization

...the memories of those who lived through those days are colorless. No trace of early morning tints, shades of blue in the sky twilight hues or moonlight silver. Everything appears covered in an ashen grey or a rotten red and brings back memories of rain and funeral rites, slime and catarrh.

—Pedro Nava on the 1918 influenza pandemic in Rio de Janeiro
(Spinney 2017, p. 55)

If not for the occurrence of improbable events, we would all be bacteria today.

—Jorge Wagensberg (Geary 2007, p. 177)

As the restrictions ebbed and flowed, I found during the openings I was not excited about going out for anything other than necessities. To go out took real effort. Perhaps this was not for the worse. The enforced solitude opened up space in my life to pursue various passions I'd neglected, like swords. I ordered several practice weapons online and they became part of my workout. I had a little-boy fascination with war which eventually folded into my peace activism. I often organized with serving military and veterans and my activist research on war became scholarship about conflict. Swords were a seductive part of my childhood militarism, especially the fantasy weapons of *Lord of the Rings* and later *Game of Thrones*, where the most ancient were usually the most powerful.

The real history of swords goes relentlessly from copper to bronze to iron to steel. Fantasies about this weapon or that helped with morale, but eventually better metals would replace weaker. Technologies evolve, not usually in a direct line but always somewhere. Swords evolved out of daggers because it was hard to make a long usable copper blade and impossible to craft one with stone. With tin added to copper, you got bronze, totally fit for purpose for long stabbing and hacking implements.

But it wasn't Darwinian evolution. Swords changed to fit the environment—metals available, military tactics, other swords and weapons—just as Jean-Baptiste de Lamarck theorized animals evolved, but it was the animal *Homo sapiens* who performed this magic consciously. It was people evolving swords, their prosthesis. This is participatory evolution.

Evolution is not just Darwinian or even the unified synthesis which folded genetics into Darwin's chance and necessity. Lamarckian processes are real, even in biology when the environment works through epigenetics to directly influence reproduction, death, and other points in the life cycles of organisms. Many have linked human tool use and human war to genetics and the relationship is hard to deny. But it is only part of the story. While evolution is real, current theories do not explain it fully. How evolution shapes consciousness and culture, and the impact of conscious choice by humans and other animals on evolution, remains unexplained.

A year into the pandemic the reality show about making weapons, *Forged in Fire*, appeared off our antenna. I'd only seen it a few times when years earlier I stayed with my parents while they were dying. They had cable. I binged it. The way the smiths interacted with their material and the history of blade making was fascinating. Human capabilities, materials available, and craft wisdom combined into extraordinary weapons. Clearly, cultural evolution is incredibly complicated.

In many ways the history of human culture is the story of technological change. In his convincing account of just how this happens, W. Brian Arthur lays out an extensive analogy to biological evolution. He starts with "all technologies are combinations" that build on earlier techs, just as life always evolves from life. Cells, for example, are the combination of once autonomous organisms as a car combines the evolutionary lines of wheels, engines, lights, seats and so on. "Each component of technology is in itself in miniature a technology," as all living systems contain other active systems. Even viruses have complicated biological subsystems for reproduction and cell capture. Finally, he notes that "all technologies harness and exploit some effect of phenomenon" from physics or chemistry. As does life (Arthur 2009, p. 23).

Integral as technology is to human existence, it isn't all of culture. Cultures evolve, but remember the most recent are not necessarily "better" than their predecessors, just more successful—and lucky. Nothing evolves inevitably to a certain end, despite social Darwinists believing white male humans were the culmination of life on Earth. This is true in culture as well as biology. David Graeber and David Wengrow show in *The Dawn of Everything* (2021) that there have been many different ways humans have chosen to organize ourselves, including giving up cities and farming, and other so-called "backward" progressions.

Cultural change is often driven by ideas, and it turns out that they are contagious. Ben Sanford Cullen even went so far as argue in his Cultural Virus Theory that viral processes were identical in both biological and nonliving things and systems. Cullen used the example of religion to illustrate ideas as parasites that could take over human bodies. Whether this is benign or horrible depends on how the host is infected, and what kind of parasite the religion or cult is (Cullen 1998).

Cullen's theory was a direct attack on the separation of biology and culture. As Michael Shanks wrote in an appreciation of his departed friend, "Ben focused on what biology and culture have in common—*processes* of maintenance and reproduction, transmission and change." The way he reconciled the evolution of the wet meat of life and the hard material markers of culture was with viruses. He rejected "the obvious distinction between the living and the dead" and proposed that "viruses are the paradigm" for evolution in both living and nonliving systems. "Culture is viral!" He proclaimed, adding:

People are living beings, *and* machine-like assemblies of working parts, *and* networked, connected ecological components. Things and artifacts are material compositions, *and* can hold agency, affecting the (human) world around them, in organic symbiosis. (Shanks 2015)

How artifacts "hold agency" is unclear, but on the complexity of the relevant systems he is certainly on point.

For many critics the obvious role of human choice in cultural change smacked more of Lamarck than was acceptable, so cultures didn't truly evolve. But it turns out that the similarities between biological and cultural change can be demonstrated by how stories change. Folklorists have shown that folk tales have long histories. "Beauty and the Beast" and "Rumpelstiltskin" are probably over 4,000 years old. The more a story is part of a group's cultural identity, the less likely it spreads to neighbors (Kuchaski 2020, p. 241). Folklore helps us understand the life cycles of conspiracy theories because their structures are the same (Ouelette 2021).

Is this connection between biological and cultural viruses an analogy or something more? Throughout their excellent *Conspiracy Theories in the Time of COVID-19,* Clare Birchall and Peter Knight wrestle with this question. Humans have no choice but to think in metaphors, analogies, and narratives. But it can be problematic. They warn us that, "The idea of escaping figuration altogether by using a scientifically objective language is naive." We "need to be alert to both the insights *and* the blind spots the different metaphors generate." Because "all explanatory models have complex figurative entanglements" (2023, p. 52).

Their concerns are articulated through how the idea of *infodemic* is used, starting with how hard it is to tell if people are using the term "metaphorically or literally, and what implications follow" from that. Most seem to use it "to suggest parallels between the way the virus spreads and the way mis- and disinformation about the virus spreads," but they worry about those who take it further (p. 45).

They disapprove of David Rothkopf's equating "infodemics to epidemics" in 2003 when he said, "In virtually every respect they behave just like any other disease, with an epidemiology of their own, identifiable symptoms, well-known carriers, even straightforward cures" (p. 45). Then they recount research that shows that the anti-vax conspiracy film *Plandemic* went viral like a virus (p. 45) and the usefulness of infodemiology techniques to analyze digital bullying in China (p. 46).

One of the reasons they worry about infodemiology is how difficult it is to figure out who are the originators of specific viral misinformation, but then they report on several cases where that is exactly what researchers did using digital "contact tracing" (pp. 46–47), and they even show how a Swedish biohacker website started the tale that Bill Gates was using the vaccine to chip everyone in the world. Among others, it infected a Jacksonville Pastor who made a YouTube video putting the idea into "Biblical language" and the rest was history, and farce (pp. 109–110). While these are exceptions, most lies in culture are hard to track to their source, the same is true of pandemics. The actual "patient zero" in biological epidemics is almost never known either.

They also have concerns that the analogy might go awry, making people think there is a "miracle cure" for misinformation, like a vaccine (p. 46). But the vast majority of vaccines aren't miracle cures, the COVID-19 vaccines are far from miraculous. People can get vaccinated and get infected, they can get Long COVID, they can die. It is the illusion that vaccines are either 100% magic or a dangerous fraud that drives most anti-vax thinking. They can't think in percentages; they can't speak Virus.

In a similar vein Birchall and Knight differentiate viruses infecting cells from misinformation infecting people because humans have agency, have "some choice" when believing and passing on dangerous nonsense. But then they go into great detail about how much of the spread of misinformation is not conscious (p. 47). Besides, who gets exposed, infected, ill, and killed by biological viruses is not entirely a matter of chance. Individual choices strongly influence who these things happen to. You can do any number of stupid things to increase your chances of getting sick and many people do.

Birchall and Knight end up equating prebunking with inoculation (p. 180) and while not saying it directly, deplatforming to quarantining (p. 181). Super-spreaders, for example, seem an integral element of all types of viral spread. On social media they are called influencers, "multiple carriers" or even "repeat offenders" (p. 46).

Still, they worry, "the turn to medical metaphors in media research risks introducing misleading analogies." The point is that many of the "recipients of conspiracy theories, misinformation and disinformation" are portrayed as "passive victims" but actually a significant number are spreaders themselves of "these forms of problematic knowledge" (p. 180). But anti-vax and anti-mask activists aren't just spreading disinformation, they are spreading the actual virus. Behavior such as this, as well as how people relate to their own health and what other precautions they take in the light of the current science, means that nobody should be a "passive victim." One can take care and still get a virus, of course, but then how do you react? Do you spread it? In the face of all contagions, biological or cultural or digital, it behooves us not to be passive.

Claiming biological contagions are completely different from cultural (including digital), ignores the complexity of the biological. Biological viruses spread, or not, in many different ways. The same is true of information whether it is shared digitally or in other channels. Yes, there are important differences between biological, digital, and cultural contagions. But there are also fundamental similarities.

Birchall and Knight complain that they can't find a specific analog for "disease" in the digital and cultural realms (p. 48), but they aren't speaking Virus. Each instance of a lie or disinformation is like an individual virus, they infect with a whole interlinked set of understandings—such as the Big Lie. The Big Lie is a disease, linked to the chronic condition of Trumpism which people with a constitutional, in both senses, weakness for authoritarianism and with racist tendencies, mainly nurtured but also linked to certain neurological presets, succumb to at much higher rates than those not so afflicted.

When they approvingly echo commentators who worry that the biology-digital-cultural framing opens up the danger of illiberal limits on free speech (p. 53) they are missing the point. Biological viruses create the same dangers, promoting undemocratic even draconian measures as in China, and unnecessary restrictions common everywhere early on in the pandemic, such as closing beaches and parks.

The concerns Birchall and Knight have for viral framing often leads to some strange comments. They report on research by the NGO Avaaz in 2020 that found "health-related misinformation attracted four times as much traffic" as information from "official health sources" did, which mirrors what many other investigations have found. But then they warn that "we need to remember" that these scientific studies start with the thesis that misinformation is spreading dangerously, just "like the conspiracy theorists they study, they are primed to find evidence for a narrative they already suspect to be true" (p. 55). Yes, that is science. All scientists have suspicions about what they will find, that is why they do the research. But there are safe guards. It is exactly *not* like irrational conspiracy theorists do their analysis. There are rules of evidence (different for various disciplines and fields), there is careful citation, peer review, blind studies, control groups when possible, and there is transparency. Irrational conspiracies depend on secrecy. Consider all the "proof" of stolen elections that somehow can't survive the light of a single courtroom procedure, even when the judges are pro-Trump. Openness is a disinfectant against misinformation in the digital and cultural realms, and a requirement in science and all scholarship.

So what is viral in the biological, digital, and cultural realms? It is much more than a metaphor, even an analogy. It is a mapping and, as this book argues, a shared language. But not the only one. Birchall and Knight point out there are other helpful framing languages:

> In contrast to the medical, economic, meteorological and military figurative language sketched out here, there is increasing interest among digital media researchers in applying ecological metaphors to information dysfunction. Ecology provides potentially productive way of thinking about the complex interactions between the content, the users, the technological infrastructure and the social dynamics of the different digital frameworks, though it is not free of its own unspoken assumptions. (p. 53)

In the end, a multiplicity of voices, heteroglossia, is the best approach. There are lots of way to tell the same stories. How a malicious story or virus changes, how it spreads, who it kills, how to stop it—these are evolutionary questions. If you don't take evolution seriously they cannot be answered.

This starts with going beyond a number of myths about our evolution that often come out as binary arguments: Nature (genetics) or Nurture? Nature or Culture? Culture or Technology? Reason or Emotion? Man or Woman? Machine or Organism? Individual or Community? When simple dichotomies are put forward to explain complex realities, they are usually wrong. Because evolution is messy, it is chancy, it is indeed black and white, and gray and blue and green and ruby and ivory and who knows what else. Evolution (one could/should say nature) is colorful. It is the monarch butterfly on the wing, even if they are going extinct.

We can't pretend that culture is somehow not natural, that humans are not really animals, that evolution does not apply to us. Most leftist intellectuals claim humans are exempt from the implications of having evolved because evolution is too slow to have made a difference in the last 10,000 years, despite overwhelming evidence that human evolution has actually accelerated in that period. They dismiss everything from sociobiology, evolutionary psychology, behavioral economics, behavioral genetics, biocriminology, neuromarketing, evolutionary biology, and neuroethics. But denial isn't the only problem.

Many scientists and journalists simplify evolution beyond reason. Genes are not selfish or evil, sperm doesn't wage wars, and technology doesn't want anything. Such claims are dangerous. This is not taking evolution seriously. To explain everything about humans with crude analogies (we are like ants, we are naked chimps, memes evolve like genes) or with simplistic (and wrong) theories—God gene, gay gene, alcoholism gene—is as much a mistake as pretending evolution doesn't exist.

It is also a mistake to use a simplistic model of evolution to pretend that *Homo sapien* is just another animal and not a profoundly unique species. Yes, every species is unique by definition, but humans have developed our own accelerated evolutionary process, and we have an impact on the global environment that is unprecedented. To take evolution seriously is to avoid all three of these errors: denial, simplification, and unexceptionalism.

Evolution always throws up viruses and viral processes, whatever the domain, as epidemiologists know well. In early 2020 a team of doctors warned that the 1918 influenza pandemic "was the deadliest event in human history" and we needed to keep it in mind as we dealt with the "growing epidemic" of COVID-19. They pointed out that the current "global, human-dominated ecosystems" inevitably leads to "the emergence and host-switching of animal viruses, especially error-prone RNA viruses." After all, "It took the genome of the human species 8 million years to evolve by 1%. Many animal RNA viruses can evolve by more than 1% in a matter of days." Unless decisive action was quickly taken, they warned, we might face the equivalent of the 1918 pandemic, or worse (Morens, Daszak and Taubenberg 2020). We know how that turned out.

Viruses and bacteria predate humans by 3.5 billion years, so when humans emerged as a new species we were already full of viruses and swarming with bacteria. While we need both types of little buggers to survive, sometimes the relationship turns deadly…for us.

Plagues in History

Athens, a charnel-house reeking to heaven and deserted even by the birds; Chinese towns cluttered up with victims silent in their agony; the convicts at Marseille piling rotting corpses into pits; the building of the Great Wall in Provence to fend off the furious plague-wind; the damp, putrefying pallets stuck in the mud floor at the Constantinople Lazar-house, where the patients were hauled up from their beds with hooks; the carnival of masked doctors at the Black Death; men and women copulating in the cemeteries of Milan; cartloads of dead bodies rumbling through London's ghoul-haunted darkness—nights and days filled always, everywhere, with the eternal cry of human pain.

— Dr. Bernard Rieux in The Plague*(Camus 1975, pp. 39-40)*

We've got to dance with the virus. There is no choice.

—*Dr. George Gao, Director of China's CDC (2017-2022) (Mackenzie 2020, p. 260)*

Plagues are inevitable history and they have changed the world's climate. Both the Justinian Plague and the Black Death led to forests growing back, resulting in significant cooling. The impact of the millions of pandemic deaths in the New World during European colonization is even clearer. That plague holocaust led to the return of millions of acres of jungle and the Little Ice Age that followed from the 16th to the 19th century (Spinney 2017, pp. 21-23). This should not surprise. We all saw wild nature rush back into the vacuum created by the initial COVID-19 lockdowns.

But the most important story is how people respond. It isn't always as horrible as Dr. Rieux, makes out. Consider Eyam, Devonshire's famous Plague Village. In 1665 the Black Death returned to London where it caused the Great Plague, killing over 100,000, 15% to 20% of the population. But in the fall it also spread to various villages, one of them was Eyam, which imported cloth from London, presumably carrying infected fleas. At least 42 villagers died before winter set in, which curtailed the epidemic's spread but did not eradicate it. Plague doesn't need human hosts, fleas feast on rats and other creatures. When the disease reappeared in people in June of 1666, the rector William Mompesson, proposed that the village quarantine itself. He was seconded by the man he'd recently replaced, the dissenter Thomas Stanley. The nearby Earl of Devonshire agreed to send food. So it was decided. The village, with a population somewhere between 400 and 800, was to suffer hundreds of more deaths before it was over. One woman, Elizabeth Hancock, buried her husband and all six of their children. Mompesson lost his wife. Altogether, at least 273 people, from 76 families, died. But they did not spread the plague to their neighbors (McKenna 2016).

In the words of a Devonshire historian from the Victorian era, William Wood:

Let all who tread the green fields of Eyam remember, with feelings of awe and veneration, that beneath their feet repose the ashes of those moral heroes, who with a sublime, heroic and unparalleled resolution gave up their lives, yea doomed themselves to pestilential death, to save the surrounding country. (Masson n.d.)

The sacrifice of this little Devonshire village is more than just a one-off story of heroism. It has lent its name to the Eyam Hypothesis, the theory that so called "sickness behaviors" (depression, fatigue, self-isolation, loss of appetite) might have culturally evolved to limit the spread of whatever was sickening us. While not proven, there is evidence beyond humans to support it, such as the mice in Switzerland who were observed maintaining social distancing when sick, a behavior that certainly spared many of their fellow mice from falling ill. While some "sickness behaviors" are a direct result of being ill, others don't seem to serve the individual's needs but those of the group. For example, the Swiss mice made ill by the scientists did avoid their fellow mice, but also the healthy mice continued to try and socialize with their sick buddies (Kaplan 2016). Humans have often done the same. In the great plagues, in the extremity of the moment, people do all sorts of things, from the most debased to the incredibly heroic. On this long list of human suffering each death is a starburst of loss and pain impacting dozens of other people. In great pandemics, grief spreads beyond the plague itself, leaving victims numb and emotionally gutted, yet in anguish. Each death counts or no deaths count.

We can ask, what if SARS-CoV-2 had spread 1,000 years ago? Or 100? 1,000 years ago it seems it would have joined in with the many other infections and afflictions our ancestors bore. Killing many, but just how many it would be impossible to say. Using 1920 state-of-the-art treatment how many more people would have died? There would be no vaccine, no oxygen machines, antibiotics, improved nutrition. The death rate would have been as high as in 1918. There are now five times as many humans as then, so perhaps 200 to 500 million dead.

Then there is Long COVID. Ancient plagues certainly left many weakened with disabilities, who died early. Just as scientific medicine has turned a large number of victims of armed conflict from fatalities to wounded, the same applies for pandemics. Outbreaks were much more deadly in the past.

Human expansion is behind many pandemics. Cholera lived in the swamps around the Bay of Bengal for millennia until the British Empire drained land for export agriculture. This unleashed six great cholera pandemics. Modern medicine, including vaccines, almost eradicated cholera around the world by the late 20th century. Experts declared it "effectively defeated" through a "triumphant containment." But it hadn't disappeared, it had "spawned a sneaky little descendant" that was "well suited to" the new environments it had colonized. It was labeled El Tor, a "supposedly mild form of cholera-like illness" but it wasn't a separate disease, it was cholera "in all its terror-inducing, virulent glory." The El Tor variant produced the 7th great cholera pandemic, the longest and most widespread of all," killing people from Nepal to Haiti, from Egypt to South Africa (Shah 2016, pp. 170-172).

Humans have a way of believing what is comfortable. Plagues are not comfortable. When they strike, reactions are strong and often irrational. Locked into their theories about humors, ancient authors could not see that diseases spread through infection. Thucydides, writing about the Athenian plague, found it inexplicably evil that both compassionate and cowardly people were struck down equally. Those who tried to help because "honor made them unsparing of themselves in their attendance at their friends house...were at last worn out by the moans of the dying, and succumbed to the force of the disaster" (1982, Book II, #51).

For Camus, however, honor was the only way forward, even if it led to one's death as it did for several of his heroes in *The Plague*. If one is likely to die anyway, why despise yourself at your end? Still, many break under the pressure of pandemic, and it has always been so. The damage can last after the contagion has burnt itself out. Irrational and selfish behavior can extend long after the dying is over. It can produce choremania, a "psychic epidemic." Laughing and fainting have spread in the same way (Jana 2022) but seldom as relentlessly as St. Vitus and St. John dances did 600 years ago.

Justus F.C. Hecker wrote about the aftermath of the Black Death, when manic infectious dancing and other bizarre collective behavior swept across the decimated lands:

The effects of the Black Death had not yet subsided, and the graves of millions of its victims were scarcely closed, when a strange delusion arose in Germany, which took possession of the minds of men, and, in spite of the divinity of our nature, hurried away body and soul into the magic circle of hellish superstition. It was a convulsion which in the most extraordinary manner infuriated the human frame, and excited the astonishment of contemporaries for more than two centuries, since which time it has never reappeared. It was called the dance of St. John or of St. Vitus, on account of the Bacchantic leaps by which it was characterised, and which gave to those affected, whilst performing their wild dance, and screaming and foaming with fury, all the appearance of persons possessed. It did not remain confined to particular localities, but was propagated by the sight of the sufferers, like a demoniacal epidemic, over the whole of Germany and the neighbouring countries to the north-west… (Hecker 1888, Chapter 1, Section 1)

Cultural responses to a biological problem is what humans do. It could be heroic sacrifice through quarantine or nursing the ill, losing one's mind in shared conspiracy theories that blame outsiders for all suffering, or collective delusions about any number of other things, trivial or profound.

Fashions and Fads, Status and Solidarity

We are a plague upon the Earth.

—David Attenborough to the Radio Times, *January 2013*

In the time of a pandemic, virus is always other people, and I *will resist any association with the term* virus. *And yet, if virus is indeed other people,* I *cannot not be a virus too, since* I *is equally "other people" in the eyes of others. In one way or another, then, we are all viruses.*

—Irvin Goh (2021, p. 146)

Around the world people responded to the spread of COVID-19 and the hard lockdowns that followed with various rituals that went viral. This seems to have been particularly common in Italy, one of the first countries hit by the pandemic. In Siena they sang the anti-fascist song "Bella Chao" and the city's song "Canto della Verbena" with the refrain, "Viva Nostra Siena." It was "Abbracciame" ("embrace me") in Napoli, "Volara" and "Grazie Roma" in Roma, opera in Firenze. Music and singing from balconies spread from house to house, street to street across the country. Individual performances became collective, became community. Banging pots and pans, clapping, and cheering for police driving by, medical people going to work, and even the garbage crews clearing away our trash, and the postal and other delivery workers bringing us things, happened around the world. These acts of gratitude never fully died out, and people continue to thank essential (and almost always poorly paid) workers to this day although not with the massed rituals that were common during lockdowns. In my neighborhood in Santa Cruz, we didn't sing nor did we send our neighbors off to work at the hospital with cheers. Instead, we howled like dogs, with dogs.

I certainly don't know who started it, and I never heard of it spreading to other cities. But soon after our lockdown started, every evening at 6, someone would start howling, soon joined by other human and canine Santa Cruzians. It lasted for weeks and was a comfort. I remember noticing when it finally died down and then ended. A lone human voice rang out at 6, but there was no answering howls. Even the dogs seem to have lost interest. The arc of the ritual felt very natural. As did the intense quiet that the lockdown brought on. Normally, I only hear the Pacific waves hitting the shore and the harbor seals barking approval at night, when the city is quiet and cool ocean winds flow inland replacing the hot air rising off the land. But during the lockdown not only could I hear more birds than ever, but even in the day I often heard surf breaking a mile away. It is all natural to me, seals and city, birds, and cars, but I do appreciate it when the balance is shifted away from the urban and suburban and even rural, to the wild. The Pacific Ocean is the largest wild left on the planet. It may be full of micro plastics and over fished and on a nice day Monterey Bay is spotted with tourists looking for whales and other thrills, but it is still wild. Humanity has not tamed it. Yet.

The category "natural" is imploding with the dying off of wildness. Across the globe wild nature uninjured or unmodified by *Homo sapiens* is going extinct. What great mountain peak or deep ocean chasm remains unvisited? There are some, and a few deep jungles, desert sands, and even Redwood forests that are free from humanity's restless making and breaking and remaking, but the air is everywhere contaminated by our emissions, the heavy metals and micro-particles of plastic, and the water is as well. The wild is dissolving through multiple intimate intertwining of the artificial into almost every macro and micro living system in the biosphere.

The cultural is now fully revealed as natural. There is no culture not evolved as nature. Human is animal, albeit a protean one. The great difference between us and all other creatures is the incredible extent we use our amazing technoscience to augment ourselves. We not only modify our bodies but the food that goes in them and the environment, the farms and cities that sustain us. Human is the modeling animal, the making and remaking animal, *homo faber*, the naked ape, the artificial ape, the civilized ("city living') ape, the cyborg primate, and now a worldwide mass extinction event. We are the transformer of the world from ecology to cybernetic organism, call it noosphere, metaman, conscious Gaia, global intelligence, or inevitable apocalypse. The speed of human evolution and biosphere domination is unbelievable, and it is accelerating. Now that wildness is vanquished, "progress" is focusing on defeating death and chance while expanding pleasure infinitely, in other words the intimate mergings of machinic and informatic systems into humans and other biological organisms and the integration of these organisms into vast technoscientific webs.

This relentless multiplying ubiquitous contagious transformation is producing a great dis-ease in our culture. As death has haunted all humanity since sentience, so now does cyborgization permeate everything we think and do. Such major transformations of human consciousness/culture have occurred before: Taming Fire, Recognizing Death and creating Story, Tools, Music, Language, Culture, Civilization and Machines were clearly on the same scale. Still, such ruptures are rare, and this current break with the past as change in our civilization goes viral, could well herald the end of humans in extinction or in the ascendancy of posthumans or machines.

But going viral is not a bad thing if you aspire to be an influencer, blogger, pundit, singer, writer, or actor. As Irving Goh puts it unfairly, this yearning to go viral "is to be narcissistic—no doubt about that. It is narcissistically driven, this viral narcissism, is terribly infectious." But it isn't necessarily just narcissism, it is often a career path or a political strategy. Being heard is not always bad, especially if it is critical and disempowered voices, from below. Goh complains "that all this is nothing less than viral anthropocentrism, in which the status of the viral has been expropriated," but he ignores the deep similarities between different types of viral phenomena. And he is just flat out wrong to claim that social media is a kind of "viral anthropocentrism, in which the status of the viral has been expropriated from its microbiological origins, with its biological lethality very much tamed" (2021, p. 146). Most viruses are not lethal to humans or even interact with us at all. Many others are necessary for our very lives, and even among those that make us ill lethality is rare. Anyway, viruses are never fully tamed, any more than dancing bears.

Goh admits that the kind of viral racism that has led to countless attacks against Asian people, and provoked #IAmNotAVirus and #JeNeSuisPasUnVirus, "can be deadly" as is "the online bullying and harassment that have driven many to suicide." After all, we must not "forget that we humans of all races, colors, and genders—we the idols of the Anthropocene—are also viruses to the world, making the world sick, poisoning its air, rivers, oceans lands, flora and fauna" (2021 p. 147).

In the end, writing in the first year of the pandemic, he embraces the solitude of mass quarantines while waiting for vaccines to be developed. He argues the best we can do "is rest, that is, to let our bodies be unplugged from all the daily preoccupations, to let the body reboot, reset itself." Because,

Chasing after viruses, whether to fanatically ride its waves (social media) or to desperately eradicate it (COVID-19) or to put a halt to its rampant pervasive spread (systemic racism), is tiring and tiresome too, after all. To rest, to reboot, to reset that is what the lockdowns…have allowed some of us to do. (p. 147)

Which has also meant not just the return of "swans and fish into the canals of Venice" while vegetation is reclaiming the land and wild animals roaming urban spaces. Social distancing and the ways it is navigated also demonstrate "that it is OK not to allow viral capitalism make every single unit of space a profit potential." In the end, Goh hopes "It is possible to (take a) break from high-speed, high-volume, high-density capitalism" (p. 148). The pandemic opened the space in many people's minds and hearts to rethink the social order, the political economy of ever increasing profits and power for an ever decreasing part of the population.

COVID-19 threw into sharp focus how trivial individual dreams of fame and fortune are, that what really makes life good is having life and health and love for each other and our communities. Love spreads, obviously. Romantic love, or at least infatuation, can even feel like a feverish infection. But love isn't limited to other people or even animals. We obsess over things, styles, rituals, as much as we do living creatures.

We all fall prey to them, some more than others. Manias for Kardashians, bell bottoms, Tide pods, Ivermectin, white supremacy, and pork pie hats are driven by our nature as social animals. Consider how yawning spreads more effectively between family members than acquaintances and, slowest of all, between strangers. Other social animals, like monkeys and wolves also have contagious yawning. Shared behaviors have to happen before they become infectious. Babies and toddlers yawn but don't catch yawning. Adam Kucharski explains, "Experiments suggest yawning doesn't become contagious until children reach about four years old." Laughing, itching, and positive or negative emotions also spread among people. Studies of teams show that a leader can infect the team with a positive or negative mood in just minutes (Kucharski 2020, p. 92).

While not as powerful an effect, the evidence is strong that smoking, divorce, loneliness, and obesity are also contagious. While the original obesity finding based on the famous long-running Framingham Heart Study has been challenged, and widely cited as well, recent research has confirmed that obesity does spread through friend networks and on to friends-of-friends. Hardly a surprise, considering the Globisity epidemic that yearly kills millions around the world (Kucharski 2020, p. 92). Now it is glucagon-like peptide-1 drugs that produce effective weight loss that are spreading like wildfire. Like songs do.

Studying a dataset of 1.4 billion song downloads from 2007 to 2014, Canadian mathematicians showed that new songs spread in patterns that closely resemble epidemic curves. Most disturbing, or not depending on your preferences, is strong evidence that different genres have different rates of infection. Electronica—listening, not dancing—is the most contagious by far, with an R rate of new song uptake at the eye-watering level of 3,430, 190 times the transmissibility of measles. Rap/hip-hop trails distantly (R=310) and then rock (R=129), pop (R=35), and metal (R=2.8). A reproduction rate of 35 is very high, actually, in the thousands is just amazing. Lead researcher Dora Roseti explains that while diseases spread through direct contact with another host, songs spread digitally. You don't even have to hear the song, just hear about it. She notes someone can tweet about a song and infect a hundred people in seconds. "You can spread a song disease far quicker than you could an infectious disease," she concludes. The specifics of any "song disease" spread depends on how it is heard or mentioned, who the audience is, and other factors, just as with biological viruses (Geddes 2021).

It turns out that gun ownership and gun violence are also contagious and so is economic behavior. But before those stories, we need to ask, why are so many aspects of culture so viral? The answer is simple, Margaret Thatcher and all the other idiots who have claimed society isn't a real thing are wrong. Society is why we are. We are social animals. Yes, each of us is unique and special and pretty precious but we are nothing without our tribe, our pack, our team, our family and friends, our culture. Spreading values and ideas between us is what makes society function. That much of this spread is viral should not be surprising.

People have been thinking about culture from the beginnings of our consciousness. Just as we wondered, and wonder still, about the night sky and the living world we inhabit, we ponder our relationship to the complicated dance that is living with other people. The digital social world, being quantifiable and stripped of many complicating factors, has become an excellent place to do this pondering about human culture, including the contagions that are constantly reshaping it.

Some of this is philosophical, even scientific, but inevitably much of it is also for profit. Early work by various "contagious labs" studying how to spread "interest" ("buzz") measured in eyeballs, clicks, links, forwards in digital life has been integral to the growth of social media. Back at the turn of the century experiments showed that the more divisive the issue the more traffic. New content was crucial for getting users to keep coming back. The chance to share ("reblogging" originally) was key for getting an idea (product, belief, prejudice) to "go viral" (Kucharski 2020, pp. 160-161).

Why would false news spread faster than accurate news? An MIT research group found that it wasn't because the false news, considered as a type of complex contagion, tended to be taken up by people with many followers, it was actually the opposite. While the spreaders had fewer followers, they were much more likely to send the false news on. The transmissibility was very high. Studies also show that in digital society the amount of exposure was directly linked to how likely it was to be promulgated. For example, on Facebook many people adopted the = sign during a campaign for gay marriage when eight of their links did. It was the same on Twitter and Skype. The more tightly linked the community the stronger the spread, but often its very closeness keeps the spread confined. Looser communities, such as QAnon, spread ideas more easily (Kucharski 2020, p. 190). It is the same with clothing styles and the latest inane teenage dance.

But fashions and fads do more than mark status and divide us from each other. They can unify in the strangest ways. Late in the first year of the pandemic, the song Jerusalema by the South African musician Master KG was turned into a dance by some of his friends in Angola, who came up with some simple infectious steps that matched the gentle, haunting melody. It went viral spreading first to Swiss police who challenged the Irish Garda, and from there to monks in Israel, on the beach in India, shoppers in French supermarkets, nuns, game wardens and their elephant friend, doctors, and airline crews (Karia 2021).

This was quite different from the dance plagues of Europe during the Black Death. Solidarity is not a mass psychogenic illnesses. One can see the difference if one watches the Swiss and Irish cops and the monks and nuns do the Jerusalema. Most of them are not dancing people, as their ironic smiles, clumsy gyrations and thought-out steps prove. But they are social people. They know society is real and they felt compelled to express their embrace of this. They know we are in this together. Not just the pandemic, all of it. Every minute of every day we are part of the long chain of being human that goes back beyond history to when we first started wondering at being alive and not being alone.

But it is not all sweetness and light. Sadly, there is something else we do together—kill.

Violence Begets Violence

Fear is an illness. If you catch it and leave it untreated, it can consume you.

—Geralt of Rivia in The Witcher *S2E1*

I have realized that we all have plague, and I have lost my peace.

—Tarrou in The Plague *(Camus 1975, p. 252)*

On June 20, 2020, in the little town of Ben Lomond nestled in the mountains above Santa Cruz, Boogaloo Boy Steven Carrillo killed Sheriff Sergeant Damon Gutzwiller and wounded a highway patrol officer before being restrained by neighbors. The same killer and a partner had assassinated a federal security guard in Oakland a few days earlier during a BLM protest, as part of the Boogaloo plan to trigger a race war.

My second son, Z, was born in our Ben Lomond cabin while I was in grad school. Santa Cruz is famous for being a very leftie—a 23% for Trump—kind of town. But the mountains are different. It is 50–50 up there or worse; some canyon roads are Aryan Nation meth industrial centers. So no surprise a Boogaloo Boy was in Ben Lomond. There are more up there for sure. This one was an active duty Air Force Security sergeant and he commuted to Travis AFB near Sacramento, so that is disconcerting.

Sgt. Carrillo and the other boys believe whites in America are being replaced by inferior peoples, the Great Replacement, and the solution is a civil war of ethnic cleansing. It isn't a new fantasy. Charles Manson was trying to start a race war when he sent his followers out to kill. For him it was clear in the lyrics of a Beatles' song. Many people live in a world where information comes encoded in obscure ways only they understand. Where, for example, black is bad and white is good. Or all cops are good or all cops are bad. While that isn't the real world people are drawn to simple viral stories. When people are willing to use violence in pursuit of fantasies real people die.

In *The Plague*, one of Camus' heroes is the mysterious traveler Tarrou, who ends up helping the narrator Dr. Rieux and others treat the sick, dying, and dead. After many days of horrors, Tarrou tells Dr. Rieux something of his life, especially how he went from being a pacifist pledged to stopping the death penalty because of an execution he witnessed, to a Communist revolutionary who was in the Hungarian Soviet Republic of 1918 when it instituted a Red Terror. After seeing "his" side execute someone through firing squad, he had an epiphany. "I came to understand that I, anyhow, had had plague through all those long years in which, paradoxically enough, I'd believed with all my soul that I was fighting it." He felt he had an indirect hand in thousands of deaths because he approved "of acts and principles which could only end that way" (p. 251).

He explains that "political" plagues are just as dangerous as biological.

[E]ach of us has the plague within him; no one, no one on earth is free from it. And I know, too, that we must keep endless watch on ourselves lest in a careless moment we breathe in somebody's face and fasten the infection on him. What's natural is the microbe. All the rest—health, integrity, purity (if you like)—is a product of the human will, of a vigilance that must never falter. The good man, the man who infects hardly anyone, is the man who has the fewest lapses of attention. And it needs tremendous will-power, a never ending tension of the mind, to avoid such lapses. (p. 253)

Our world is a continuation of the 1930s that shaped Camus. Violence is everywhere.

Violence epidemic is not just a figure of speech. Violence (murders, assaults, accidental deaths, and suicides) spreads like a disease. Dr. Gary Slutkin, an epidemiologist, notes the best predictor for violence is "a preceding violent event." After years of fighting diseases in Africa, Dr. Slutkin returned to the United States and realized that the deadly gun violence that was killing thousands of people there was also an epidemic, sharing the same life cycle and other aspects of biological diseases (Kucharski 2020, pp. 120–129).

Charlotte Watts, another epidemiologist and an HIV expert, showed in the 1990s that the propagation of domestic violence also mirrored biological spread. There was the importance of previous exposure both as victim and perpetuator, "dose-response effects" (the more violence one is exposed to the more one will be exposed to), the clustering of outbreaks and the central role of super-spreaders. A study in Chicago showed the R of a gang killing was 0.62 (p. 125). In the U.S. it is shootings, in the UK knife stabbings spread in a very similar way (p. 129).

Suicides spread the same. They are often sparked by earlier suicides and explode in clusters, but this hasn't made profit-driven mass media more careful. There was a 10% uptick in suicides after Robin Williams hanged himself, with a particular rise in middle aged men killing themselves through asphyxia due to hanging (p. 123).

While these generalities are telling, the devil is in the details. Carrillo and his fellow Boogaloo boys are trying to start a violence epidemic. In radical politics this goes by the name "accelerationism," and it is advocated by both "left" and "right" extremists, however vague those categories are. "Heighten the contradictions" is the idea in Marxist lingo. "Bring the War Home" was the Weathermen version of this. Revolutionaries are looking for that spark that sets off the fire of war. Revolution is war, after all, starting with the struggle for the hearts and minds of the people. These days, one often gets there digitally, so soldiers and agitators alike talk about hybrid war, low-intensity conflict, imaginary war, cyberwar, sim war, net war, information and info war, neocortical war, sixth generation war, knowledge-based warfare, net centric or network centric war, mimetic or meme war, cybernetic war, cognitive warfare, spectacle war, and internet war. All wars, real conflict, none-the-less (Gray 2005, pp. 24–25).

There is a growing focus on social media as a battleground. Radical groups and established militaries are hunting for ways to weaponize the algorithms and other processes Facebook, Google, Twitter/X, and others use to maximize their profits and control. New technologies create new relationships between violence and power.

Power is the ability to do something. Power cannot exist in and of itself or in a vacuum. It has to be useful. All power does not corrupt. Certain kinds of authoritarian coercive power are horrible, and Lord Acton's maxim certainly applies to them. And as absolute coercive power becomes technologically possible, we face the danger of absolute authority and corruption (Gray 2001). But much power is constructive and some power is delightful, even edifying. Consider child-rearing. Having loving relationships is a powerful act that changes the children one is responsible for in often positive, even wonderful, ways. Power can be subtle, it need not be obvious. Perhaps this is one of our key political problems. Most types of power that have been well analyzed are violent and unsubtle, such as military action or interpersonal violence.

The realm of legitimate deadly violence is war, a venerable institution. The contagious elements of war, traceable back to prehistoric ritual war, are now being turbo charged by military revolutions in information technology, the shift to the digital, which is also organized around infection and, inevitably, exponential escalation, destruction, and collapse (Gray 1997).

Directly or indirectly, humans cause most premature human deaths. Not only is there the day to day violence of criminals and police, often the same people, there is domestic violence, there is war, there is revolution. There are incredible inequalities leadings to starvation, malnutrition, illness, and the other unnecessary deaths in the viral underclass.

Obviously, violence isn't always catching. As with any disease, infection rates vary by all sorts of factors. But then again, in American culture it seems it is encouraged—If it bleeds it leads. If it is hot it sells. If you want a gun you can get one. If you are angry, you click more...

Violence, whether physical or emotional, is about imposing one's worldview on someone else. Elaine Scary's eloquent book about torture and war, *The Body in Pain* (1987), shows this clearly. Violence is a relationship. That is why most of it happens between intimates. But violence happens between strangers as well, sometimes encouraged by other strangers.

Shoshana Zuboff explains that making the Interweb an instrument to manipulate behavior is not just about selling things, it is also used politically to motivate affiliations, voting, and even violence. "Competitive production, profit maximization, productivity, and growth" are the new logics of accumulation "defined by extraction and prediction" of human behavior. This leads to a...

...means of behavioral modification that incorporate its machine-intelligence-based 'means of production' in a more complex system of action and the ways in which the requirements of behavioral modification orient all operations toward totalities of information and control, creating the framework for an unprecedented *instrumentarian power* and its societal implications. (2019, p. 67)

This is an actual conspiracy which spreads irrational conspiracies. As Corey Doctorow put it in May of 2020, "Why is it so easy to find people who want to believe in conspiracies? My answer: Because so many of the things that have traumatized so many people ARE conspiracies." He goes on to write about the Sackler family buying off regulators to foster the deadly opioid pandemic, Boeing pushing 737s into operation even after they started crashing, politicians using pandemic briefings to buy and sell stocks, and so on. His conclusion is undeniable: "In a world of constant real conspiracy scandals that destroy lives and the planet, conspiracy theories take on real explanatory power" (Guffey 2022, p. 22).

5

Political Contagions

The Rise of Algorithmic Intelligence

If it works, it isn't AI.

—*Joke from the 1980s. Somehow, not funny these days.*

...the rise of an algorithmic society in its current form anticipates the decline of democracy—of the belief that one shares a stake in a public good, that society is governed by fairness, and that dissent can be effective. While political elections may continue to be the visible mark of democratic practice, the fight for democracy in an algorithmic society will be about accountability and transparency for those who control the software and access to data, how consent is reimagined and protected, and how challenges to injustice are protected and supported.

—Catherine Besteman (2019, p. 180)

In the summer of 2022, as Omicron 5 spread across the world, I watched the hearings of the January 6 Congressional committee. It was almost as intense as seeing the coup attempt live toward the end of the Delta wave. I've helped take over a few places in my activist life—several offices and the Old Union at Stanford, squats in three countries, a utility headquarters in Sacramento, part of the Seabrook construction site, a courthouse in New Hampshire, a ridge overlooking an ICBM test at Vandenberg Air Force Base, and the California Capitol. I have been at a half dozen protests at the U.S. Capitol, including several that were pretty much police riots against Vietnam Veterans Against the War and their allies. Imagine my surprise at how easy it was for the MAGA mob to storm inside.

But that coup failed. As of this writing (winter of 2025), it is unclear as to how Trump's second coup will fare. Using Elon Musk and his DOGE teams, it is a new kind of seizure of power, using "control" of "software and access to data" to render democratic consent irrelevant, just as Catherine Besterman warned. As she pointed out, "the fight for democracy in an algorithmic society will be about accountability and transparency" and will necessitate a reimagining of consent and "how challenges to injustice are protected and supported."

The Congressional hearings showed how Trump mobilized his tatty army to disrupt the certification of the election, and perhaps "hang Mike Pence" or even worse. Trump did this with two types of viruses—viral lies and viral digital techs to spread them. The same strategy that got him re-elected in 2024, but the second time around the overthrow of democracy is being attempted with algorithms and AI.

It is a commonplace among tech pundits that relentless and accelerating improvements in technologies of all types is speeding up cultural and biological changes, with a major factor recently being the algorithms of viral AI (Crichton 2021). Since the 2008 Financial Crisis, technology corporations have developed forms of value extraction and capital accumulation which depend on a new variant of AI. This system is called digital, platform, or surveillance capitalism. According to Javier de Rivera (2020), the strong performance of GAFAM (Google, Apple, Facebook/Meta, Amazon, and Microsoft) is the result of a structural alliance between financial and technological elites, where the technology sectors are dominant. The role of these corporations in our lives is normal now, and so is the increasingly common resignation with which we accept a state of permanent surveillance and manipulation through our computers and phones.

In addition to increasing sales through the knowledge of users' lives, this "behavioral surplus" is leveraged politically to influence voting in elections, foster political discord, and incite people into the streets for marches, rallies, and even violent confrontations. Shoshana Zuboff (2019) demonstrates in *Surveillance Capitalism*, this instrumentalization of the Interweb is now one of the most powerful ways of manipulating attitudes and behaviors in the world. It is used to sell us music, politicians, books, games, and so much more. No wonder surveillance capitalism is so scary, it is everywhere in our lives. And it runs on AI.

What AI means is always being redefined by what can be done technologically, how the label sells, and the tension between the views of computerists, philosophers, and popular culture. The seemingly sentient systems that Alan Turing and countless SF (science fiction or speculative fiction, take your pick) writers dreamt of is today called General Purpose

Intelligence (GPI, also "hard" AI, also "our future masters"). The AI we are getting now, based on large language models (LLM), is called generative/viral AI or, my preference, Algorithmic Intelligence, coined by my colleague, Prof. Ángel Gordo of Universidad de Complutense.

When AIs learn from big data sets, be it face matching or art making, they are like viruses evolving: (1) take many tries; (2) save the ones that are closer to working; and (3) many more tries. Think of the individual algorithm iterations as individual viruses. When humans hijack biological viruses to fight bacteria or make vaccines they have a goal; engineered digital viruses also are always part of a larger plan. For example, facial recognition programs modify themselves by running analysis of faces over and over again and scoring its efforts. As Kayla Ismail (2018) points out, unlike more limited algorithms, "data is created, transformed and moved without [...] any operator or agent."

AI today depends on masses of simple calculators, as a viral wave is made of individual viruses. Algorithms are no more than little hunks of semi-intelligence, a recipe or command sequence producing some specific result and then doing it again. AI algorithms make the social media networks function, and their primary goal is more profits from surveillance capitalism (Zuboff 2019; Gray 2019; Masco 2019). As a side effect, they facilitate ("afford") the spread of viral ideas, such as eating laundry detergent and QAnon.

Now we not only have impressive AI art, text, and coding but also AI-powered viral marketing, contagious buying, and other aspects of surveillance capitalism producing algorithmic governance (AI making decisions for police, social workers, and the military), all with back propagation driven by viral dynamics from evolutionary programming, seeking out exponential growth—which means more hosts/growth mediums. This is why the military has long invested in AI, looking for useful variants to help dominate the info sphere that is a major front in postmodern conflicts, as the Ukraine War and the Torture of Gaza prove (Gray 2025).

Both military AI and the e-economy depend on incredibly large amounts of reiterative data processing. Whether it is target recognition, regulating treatments of patient-customers for profitability, or matching customer-students to for-profit online education platforms, AI driven roboprocesses are crucial, and dangerous. Robosigning of fraudulent mortgages produced the 2008 financial crisis in the U.S. (Stout 2019). Roboeducation systems configured for maximum profitability are fundamentally changing teaching and learning (Fernandez and Lutz 2019). In the U.S. automated sentencing and other legal determinations are degrading an already flawed system (Middlemass 2019), and the same is happening in immigration decisions (Terrio 2019).

Kai-Fu Lee argues one of the most important challenges for AI is unsupervised learning from raw data without human intervention because it often produces racist programs that, for example, penalize African-American names in resumes and discriminate against people of color in general. The data sets, and assumptions, that are used for machine learning will reify the racism (and other hatreds) of society unless great care is taken (Lee 2018). A related danger is the deployment of emotional roboprocesses. While computers that simulate emotional relationships might produce "lives full of love," it is more likely that they will generate "bizarre outcomes that utterly fail or a hallowing out of human emotional life" (Gehl 2019, p. 116).

Hugh Gusterson, co-editor of the dire collection *Life by Algorithms: How Roboprocesses Are Remaking Our World* (2019), summarizes their impact in six dynamics: disciplining and deskilling, income extraction and commodification, growth of unaccountable wealth and power, the standardization of the nonstandard, remaking people and relationships, and an inevitable response producing new "strategies of agency and action" (2019, pp. 13–23). Gusterson summarizes,

Thus the conjuncture between computerization and neoliberalism has produced roboprocesses skewed in favor of corporate profit making, mass surveillance, and the retrenchment of racial and class-based inequalities. (2019, p. 7)

Roboprocesses tell us what to buy, how to get to the sites we want, which is the cheapest hotel, who might be our ideal partner and "if we are eligible for a loan or who to vote for." Advances in machine learning have created algorithms that allow corporations and governments to implement new forms of monitoring and manipulation, establishing algorithmic governance in various parts of society. They push the functions of the State that guarantee social order to the background (Sastre and Gordy 2019, p. 160) and leave the exploitation of our behaviors and preferences in the hands of the large digital oligopolies, unless the State is driving the process, as in China. Antoinette Rouvroy and Thomas Berns (2013) argue that this is a new form of algorithmic governmentally that threatens to extinguish actual identity in digital simulations (see also Fielitz and Marcks 2020). When the State and oligarchs become one and the same, as in the second Trump administration, the danger to democracy becomes extreme.

In response to this danger, there have been a number of new projects to keep track of the growing power of algorithms and think about how to control them, especially since Google and other behemoths have rejected any serious critiques by continually purging their own ethics teams (Grant 2022). A Berlin NGO, Algorithm Watch, has come up with a useful "AI Ethics Guidelines Global Inventory" which they regularly revise (Haas and Gießler 2020). Data & Society, in New York City, has also put forward AI guidelines as well as a proposal for Algorithm Impact Statements, modeled on the plethora of environmental and similar impact statements. In general, Europe has taken a much harder line on AI and algorithm driven applications with much stronger regulations than the U.S. This is reflected in the proposals of the two think tanks.

Data & Society is naive about the reality of regulatory capture and the role of experts in the procedures they propose. With enough money one can buy whatever impact statement one wants. Unsurprisingly, their AI guidelines were taken up by the Biden Administration under the rubric of an "AI Bill of Rights" (Haven 2022). It consisted of vague unenforceable prescriptions for niceness and will only apply as advice to the U.S. government. It doesn't even pretend to be relevant for the private sector. A sign of its general uselessness is it appears they don't realize we don't need a bill of rights *for* AI but to protect us *from* AI. The second Trump administration withdrew it anyway.

The regulatory approach of many NGOs to lobby government for liberal policies of prevention and amelioration ignores that the fundamental issues are the power and the profits that prevent real regulation, and the reality that AI is improving exponentially. Superficial responses will fail miserably. We have seen this in the rise of social media.

Now there is a new form of authoritarian populist cyberactivism based on gamification. It's not that sophisticated games can't empower people to defend the environment, for example. But new gaming techs and theories also afford reactionary mobilizing. QAnon, in particular, is based on the gamification of right-wing—especially white and male—alienation, driving decentralized virtual recruitment and self-activation. It is powered by the viral algorithms of the social media companies that amplify negative and extreme emotions to increase the stickiness, and profits, of their platforms. None is greedier then Facebook.

The Politics of Social Media

Reactionary concepts plus revolutionary emotion result in Fascist mentality.

—Wilhelm Reich, The Mass Psychology of Fascism

Social media is the game playing the players.

—C. Thi Nguyen (2020)

The first person I ever blocked on Facebook was the first person I knew who predicted Trump would become president. At the time I thought that was impossible, and that belief only became stronger as he was revealed to be a pussy grabbing, pro-Russian, incompetent racist grifter. But I was wrong and my friend was right. Still, I unfriended him. Why? Because I thought he was part of the problem, not the solution.

An accomplished artist-provocateur, during 2016 my former student and friend, RW clowned (literally) presidential candidates, including Trump. While reporting on Facebook about his in-person experiences at campaign events, he also reposted many memes, especially attacking Hilary Clinton, that we now know were produced by Russia. RW and I are far to the left of Clinton, so I certainly don't mind criticisms of her horrible foreign policy or general 1%-style corruption, but crap about murdered staffers and Benghazi betrayals just detracts from her real problems. Besides, compared to Trump she was the love child of Abe Lincoln and Emma Goldman. So I would debunk the worst of the memes and argue he should not repost obviously fake information. He responded that he'd repost whatever he thought was interesting, and his followers would have to sort it out.

Finally, he replied to one of my critiques that he was considering unfriending me because of my constant criticism—also known as fact-checking. So I unfriended him. It felt good, but no doubt he also felt relief from my carping. Since then I have had to block a few other old friends. Sometimes, the psychic pain of staying engaged to people whose worldview is relentlessly irresponsible or even demonstrably unhinged, is just too much, even if we share history.

So why is there Facebook? Profits? Yes, but more than that. Zuckerberg used to end meetings with a simple declaration—"Domination!" Power is what Facebook has been about from its beginnings objectifying (rating), and then targeting, new Harvard coeds. Mark Zuckerberg's goal, as with Bezos and Gates and Musk, is much more than mere money. It starts with capturing your attention; it moves to owning your data; it ends with controlling you. At the 2024 Meta Connect conference, Zuckerberg wore an *Aut Zuck Aut Nihil* t-shirt, in homage to the original, a slogan of the first God-Emperor of Rome: "Either Caesar or Nothing" (Hyde 2024). We can't say we weren't warned; he needs to remember that every Caesar has their Brutus.

Kate Losse was Zuckerberg's speechwriter until she quit because of the Cambridge Analytica ("Data-driven behavior change") scandal around Facebook mining user data. Officially, the company professed shock, but she says it was common knowledge that they sold their data to third parties. It didn't help that people who raised concerns about such practices, selling private data to right-wing companies to manipulate elections, were like "wife beaters and suicide bombers" in Meta's view (Losse 2018).

The company's business model is based on fraud. It is a consistent lawbreaker and "profoundly evil" (Dayen 2021). Besides that, it is a danger to "the economy, health, and democracy" (Sitaramana 2019), doesn't care about fact-checking and is a threat to journalism (Levin 2018), knows it is radicalizing extremists (Zadronzny 2021), makes money off climate lies (Gilbert 2021) and white supremacy (Tech Transparency Project 2022), and even fought a California law to ban addictive apps for children (Belanger 2022). Since ending most of its efforts to monitor anti-social material, Facebook has implemented algorithms that encourage it. For example, in 2024 it didn't just allow right-wing militias to recruit and coordinate trainings, it automatically generated pages for them to use (Owen 2024).

Some scientists think social networks are like infectious diseases (Kerr 2014). In 2014 some academics modeled Facebook as a pandemic and predicted that by 2017 it would lose 80% of its users. *Guardian* columnist Arwa Mahdawi (2014) explains clearly:

The Princeton researchers make their case via epidemiological modeling, acronyms, and lots of formulae where the γI terms in equations 1b and 1c are multiplied by R/N to give equations 3b and 3c. Quite frankly, this means the sum total of F/U+C*K all to me.

Actually, Facebook went from 1.3 billion users in 2014 to 2 billion in 2017, up 50%. At the end of 2021, it had almost 3 billion. Still, that was the same as six months earlier, the first time ever it didn't grow. So maybe the model isn't totally off. Facebook does have its limits, both Threads and the Metaverse have failed.

Mahdawi's overall point is that social media is often infectious and toxic, we don't need formulas to know it. Just because some scientists got the R/N off doesn't mean it isn't true because the symptoms are all around us, even in us. Facebook Fever—*Harvardus aluminus* she calls it—has infected all its users, although many have recovered. It infects to unfunny effects. When President Biden said that by allowing vaccine misinformation to spread social media such as Facebook was "killing people," many infectious disease experts agreed (Lieser 2021).

Meta aims to harness augmented reality and alternative reality gaming approaches to monetize even more behavior. They want to turbocharge Facebook's current approach with predictive algorithms driven by massive data mining to manipulate people with "emotionally contagious" content, as they first demonstrated in 2012 when they changed users' content to drive positive or negative reactions (Rose 2021). They have learned how to manipulate people to an extraordinary degree.

So how does one direct right-wing activists to disrupt voting or storm centers to stop the counting? Tell them it is to save the election, to defend democracy. They will hear through friends, on 4chan, X, Facebook, Nextdoor, email or whatever, that illegal immigrants are being bused from precinct to precinct in Maricopa country (60% of Arizona's vote) and patriots need to go to the polls and challenge every voter who looks illegal. Or the ballot counting in Miami and Scranton is slow because the Democrats are changing votes. Time for patriots to go and take over the process.

We know how these stories spread. In June of 2020 bands of scared, well-armed, white men swarmed locked down Arizona malls and many other places across the U.S. because they heard reports of immigrants storming suburban houses and raping women. These stories were traced to Nextdoor postings that were based on tweets from Willow Aldridge, who does not exist. It is a fake account to amplify messages (Anglen, Rueles and Long 2020).

The wave of "Antifa-lit-the-fire" rumors during the inferno summer of 2020 in Oregon led to different gangs of frightened and armed white men stopping reporters and others on backroads in the fire zones around Medford, Oregon. These stories were spread by Donald Trump supporters such as Paul Romero, a local GOP candidate (Markey 2021).

Unsurprisingly, QAnon nets also spread these arson accusations (O'Sullivan and Toropin 2020). Another amplifier of false reports about Antifa fires, climate change and BLM protesters was the 78 Federal-State Fusion centers, often co-located with FBI Joint Terrorism Task Forces, across the country. Sharing officers from state and local jurisdictions with at least one Department of Homeland Security official, they are supposed to combine intelligence from all sources (CIA, NSA, and the U.S. military) to alert law enforcement to domestic threats and co-ordinate the response. But as the Blue Leaks documents prove, they can be quite susceptible to wild theories about left-wing protesters (Devereaux 2020). Ironic, as one of their main tasks is rumor control. It was proudly fascist Identity Evropa that used fake (@Antifa_US) accounts on Twitter, amplified by some Fusion Centers, to claim Black Lives Matter was about to invade numerous white suburbs (Kates 2020). This set off a panic in various white enclaves across the land (Portland, Minnesota, Denver) leading to groups of...well, you can guess the rest.

Cyber operations in Lithuania, Georgia, Ukraine, and the United States have allowed Russian government hackers to hone their skills in spreading such rumors. Russia has encouraged racist violence in every U.S. election since 2016. Former CIA official Ned Price explains Russia's twin goals were to stoke chaos and elect Trump. "They are essentially one and the same" (Derysh 2020).

But we don't need Russia to spread hate. It was Kevin Mathewson, a former Republican alderman in Kenosha, who set up the Kenosha Guard

Facebook site (Shortell, Carrega, and Campbell 2020). A 17-year-old in a nearby state saw the post and came to Kenosha to "protect" businesses from BLM protesters and killed two white anti-racist men. The acquitted killer is now hailed as a hero by the whole right-wing spectrum, including most Republicans. He even has a video game.

Psychologically, the "man on the white horse" is necessary for reactionaries, but with virtual recruitment he is really only symbolic. The actual movement is articulated and expanded on the Interweb through cryptic messages and a logic of riddles and games. Some angry and confused people seek to "solve" their distress by uncovering (with the "help" of QAnon moderators) the truth about the massive pedophile conspiracy led by the coastal elites (that they are sure despise them) and the subgroups they fear—Communist Jews, Blacks, Asians, Latinos, immigrants, uppity women, scientists, and liberals. Through the gamification of their alienation they become true believers. So what is behind their post-truth, fake news, negation of democracy?

Post-truths are not easily moldable or correctable. Their dependence on beliefs with a strong emotional charge makes them resistant to evidence and fuels their rejection of any information ("fake news") that conflicts with their world view. This means personal emotions and stories are more influential for much of the public than validated facts, science, or reasonable arguments (Fasce 2020). In essence, "the problem posed by post-truth is not a mere stain on the mirror. The problem is that the mirror is a window to an alternative reality" (Lewandowsky, Ecker, and Cook 2017, p. 356).

Post-truth beliefs corrupt political decision-making, such as Trump's policies around the coronavirus: rejection of masks and social distancing, belief in magical cures, denial, and vaccine conspiracies (Borreguero 2020). These claims are integrated into Trump's Big Lie about the election of 2020 and the assault on the Capitol. He claims he was "stabbed in the back" and tens of millions of Republicans believe him. The similarities to Hitler's Big Lie about Germany's defeat in WWI are not a coincidence. While every nation-state has its own story of Authoritarian Populism, they share many elements. Lately, thanks to the global Interweb, conspiracies can spread faster than ever. What Polly Price said about biological contagions, seems relevant:

America's experience with plagues past and present shows that effective disease control is not just a public health task of containing pathogens, but also a summons to thoughtful law and reasoned governance. History proves that democratic governments like ours have inherent weaknesses in their ability to control epidemics, especially if coercive measures become the last line of defense. (2022, p. 203)

Successful conspiracies are infectious. Many of their viral elements are inherited from older conspiracies and accelerated by the latest techs. They are also conscious acts of invention. QAnon's key fantasies can be traced back several thousand years, but what makes it particularly virulent in the 21st century are its digital augmented reality (AR) live action role-playing aspects. You join QAnon by playing ("Do your own research!") but its goal is not fun nor even the truth. It is taking power.

QAnon and the Victory of MAGA

QAnon is the game that plays the players.

—Reed Berkowitz (2020)

But where some saw abstraction, others saw the truth.

—Camus (The Plague, 1975, p. 92)

Trump's 2024 victory, and the Musk/DOGE coup that followed it, can only be understood if one speaks Virus. It took literally billions of postings on X and hundreds of millions of dollars of social and other media expenditures for a majority of voters to give Trump credit for Obama's economy, and Biden/Harris the blame for Trump's. Trump's inflation, along with the greedflation of most major consumer companies, made it feel to millions of Americans that the conman with multiple bankruptcies was the best president for the economy. Racism and misogyny were key factors, but they needed to be inflamed by viral anti-immigrant and openly women-hating messaging from Trump and his allies in the manosphere—AI digitally spreading a cultural virus of hate and fear resulting in our future. Or at least one dark important strand of it. It was an engineered political pandemic that produced regime change. This was probably why QAnon was created.

QAnon posts first appeared in 2017 from someone claiming to be a government bureaucrat with top level (Q) State Department security clearance. They were helping President Trump take down a global network of sex traffickers led by Barack Obama, Hillary Clinton, and George Soros, among others. This Dark State of senior officials and former political leaders plans, or has already staged, a coup but in many versions of this story Donald Trump, secretly still president until 2024 (nonsensical since the 2024 election), will soon launch a "Perfect Storm" to crush the Dark State and trigger a "great awakening" to save Christian White America (Jerez 2020; Thompson 2020).

A minority view in QAnon is that this worldwide conspiracy is controlled by lizard aliens, until recently led by Queen Elizabeth II. There are other theories thriving among the motley alliance of COVID and vaccine deniers, enemies of 5G and more traditional fascist, nationalist, and white supremacist activists. Andreu Jerez (2020) points out that in QAnon, "everyone agrees to maintain an attitude of great mistrust towards governments" even as they don't look too closely at the beliefs of other Qs, and count on Trump, elements of the military, and right wing politicians to carry out the Perfect Storm.

Lizard queens are great fantasy but my favorite story from the Q-verse is adrenochrome. The hormone adrenaline is produced by the adrenal glands, oxidize it and you get adrenochrome. Hunter Thompson claimed it was a popular drug in *Fear and Loathing in Las Vegas*. Among his many humorous assertions is that it is only available through harvesting from live humans. Not true, scientists synthesize it easily. Another Thompson detail is that it is a powerful ritualistic high. In real life it is being scared shitless, often while hallucinating. Somehow, probably as an inside joke ("You think they'll believe this?") by QAnon's instigators, pedophiles harvesting adrenochrome became a central point of QAnon doctrine. Among the places it was obtained by torturing children was the nonexistent basement of Comet Ping-Pong Pizzeria in Washington, DC, which one Q seized at gun point, shooting open a closet looking for captured children (Lee 2022). QAnons have found more "proof" of this evil drug cult in statues and plaques from the Middle Ages showing the apocryphal murder of St. Simon of Trent by Jews, who, so the lie went, needed the blood of Christian children to make Passover matzah. This is the "blood libel" used by antisemites to launch pogroms and other attacks on Jews from then to today (Lee 2022).

But QAnon is not privy to arcane medieval wisdom. The actual origins of QAnon are clearly connected to the PSYOP community of the U.S. Army. A key player is Major General Paul Vallely (Ret.). A veteran of two combat tours in Vietnam, he served around the world, including commanding all Special Forces, Psychological Warfare and Civil Affairs Units in the Western U.S. from 1982 to 1986. On October 14, 2019 he claimed in an interview that Q was real and got information from "The Army of Northern Virginia" ("800 military and civilian security experts"). Supposedly, Trump listened to them because he didn't trust the CIA or DIA. Gen. Vallely was the most important military figure to embrace QAnon, until Trump openly did on July 1, 2024, right before he was elected Commander-in-Chief, "retruthing" a picture of himself and Melania in formal dress with the Q rallying cry: "Where we go one, we go all" (Nicholson 2024). Not a stupid move, QAnon's support had been climbing since Trump lost in 2020, going from 14% in 2021 to 23% of all Americans in 2023–including 29% of Republicans. 30% of White evangelical Protestants, and 26% of all Black Protestants. Support from Democrats doubled to 14% (PRRI 2023).

Vallely was co-author of the infamous military memo "From PSYOP to MindWar" (Valley (sic) and Aquino 1980) when he was commanding the 7th Psychological Operations Group. QAnon devotees make much of this memo, claiming it is the plan for the child trafficking cabal they struggle against. They point out Vallely's co-author was Major Michael Aquino, an active Satanist and founder of the Temple of Set, accused (but cleared) of raping a young girl. They don't mention that General Vallely has validated QAnon, nor that the memo actually describes something much more like QAnon than the supposed child torturing cult of their enemies (Guffey 2022, pp. 75–78).

The essay is remarkable because it was the first articulation of the psychological power of deceptive immersion—total rapport between manipulator and the manipulated—that has only become possible on a large scale since the spread of the social media applications on the Interweb. Through Reddit, 8kun, Twitter/X, Facebook and other sites, people looking for the kind of truth QAnon offers can easily find it. Robert Guffrey notes this is crucial: "For the mind to believe its own decisions, it must feel that it made those decisions without coercion" (Guffey 2022, p. 79).

Retired military PSYOPs operatives worked with right-wing hackers and far-right political operatives like Jim Watkins (former owner of 8kun, QAnon's main platform at one point), Brad Pascale (Trump's 2016 social media campaign manager, fired in July 2020 for not gathering enough people at a rally), Roger Stone (Trump adviser, convicted of lying about Russia's attempts to boost Trump's presidential campaign in 2016, pardoned by Trump), retired Lt. General Michael Flynn (Trump's first National Security Advisor, also pardoned by Trump), Steve Bannon (Breitbart; Cambridge Analytica; Trump campaigns, administration and coup attempts, pardoned by Trump) and many others (Guffey 2020).

In the 2024 campaign Trump's links to QAnon became even closer. RFK Jr.'s submission to Trump is a prime example. QAnon ideas have become mainstream now. Jesselyn Cook reports, "They're repeated by major political influencers, right-wing media stars and elected officials, and they appeal even to everyday people who would never consider themselves to be affiliated with QAnon." The Kennedys have long been a QAnon fascination, with various subsets of the cult believing one of the dead ones was about to return. RFK Jr. doesn't just scratch this QAnon itch, he brings in the anti-vax crowd as well. When BGatesIsaPyscho, a leading QAnon influencer with half a million followers, posted about the danger of chem trails, Kennedy promised him "We're going to stop this crime" (Marcotte 2024).

Kennedy came into power along with Trump, as did Elon Musk, who on election day sent a supercut video to everyone on X (he owns it after all) that was pure QAnon celebration of Trump (Oamek 2024). Considering Musk's companies: the Starlink satellite-internet constellation, X, Tesla. and SpaceX, he is a formidable ally for Trump. But for all their skill in mobilizing angry and fearful voters, Trump/Kennedy/Musk did not make them that way.

A big part of the appeal of QAnon is that the suffering of working people under the current economic system of extreme wealth concentration is openly acknowledged. In the Q Game it is assumed things are very fucked up. Unfortunately, solving for Q means believing in an authoritarian cartoon version of reality and accepting a racist, antisemitic and misogynist world view that rejects science and claims the only way to save

American democracy is to destroy it. The power of Q comes from anger against the rich that is turned into hatred for other people, while becoming part of a community of like-minded believers. They have their own language with slang (QAmoms), and slogans: "problem, reaction, solution" and "do your own research." There are mottos "where we go one we go all" (WWGOWGA) and shared group paranoias about phantom catastrophes such as The Great Replacement and The Great Reset.

But QAnon's emergence was only possible once social media companies took over the Interweb. At its birth, military and other hierarchical institutions could not control it, despite their sophisticated use of weaponized information (intelligence, propaganda), and being the original promoter of cyberspace. But it was non-hierarchical by design. Much of the success of the social movements that emerged in the last two decades, from the Arab Spring (2010), to 15M (2011), to Occupy (2011), to Idle No More (2012), was due to decentralized decision-making and mobilizations facilitated through the Interweb (Gray and Gordo 2014).

But machine-learning AI fundamentally changed the culture/coding of our digital reality, starting with the colonization of most of the Interweb by social media companies. Authoritarian actors, such as the Russian and Chinese governments, have also mobilized these new technologies effectively. The decentralized logics and autonomous decisions of leftist movements were far from being their exclusive characteristics. On the contrary, the Interweb's own logic, its non-hierarchical and decentralized organization, can also be manipulated by rightist networks, as QAnon proves. It is a new kind of politics.

Using negationist logics ("if it doesn't match our story it is fake"), reactionary groups have promoted a new form of "self" activation. Q doctrine is based on 5,000 or so (Q adherents can't agree) Q drops, enigmatic statements released with specific trip codes—hashed passwords which allow postings across threads on 4chan and 8kun, controllable by systems administrators. These appear nonsensical but some are encoded in puzzle format with clues seeded on photo boards of the 8kun portal. The rest is just made up out of nothing, apparently. The interpreters of the Q droppings are the real leaders—influencers—in the Q world. After Q messages ceased (December 8, 2020) these moderators became more important than ever.

"QAnon is like a game, a dangerous game" Clive Thompson (2020) declared. This makes it very addictive. Belonging to QAnon, argues Thompson, means being part of an Alternate Reality Game (ARG) where you work with thousands of people to decipher a real life mystery and *save the children!*

ARGs blur the lines between physical and virtual realities. Steven Jones (2009) explains, "the players interact in a space that overlaps with the real world; physical spaces and objects are coded with geographic locations and this allows dynamic and interactive interaction situations" (Díaz 2017, p. 29). Unlike augmented reality with transmedia narratives, such as *Star Wars* or *Pokemon*, in pure ARGs the players themselves tell the story on the streets and on different platforms, creating an experience where fiction acquires "a certain life" (Díaz 2017).

QAnon blurs reality and fiction by encouraging the players to transcend the narrative of the game itself. Thompson (2020) details how it worked: Q (systems administrators pretending to be Q) leave Q drops on a website. Followers decipher them, although in reality only a few moderators did much research. They "discover" new clues in coincidences and the creative interpretation of emails and other data from Deep State suspects. Inevitably, this "analysis" supports right wing positions in general, and specifically the more outrageous claims of Trump, as well as the core story of Deep State pedophiles about to be arrested and executed.

Game designer Reed Berkowitz (2020) notes that QAnon is "played" while it re/creates reality according to the logic and interpretations of the Interweb. Thompson (2020) maintains that "ARG and QAnon represent the quintessence of Internet culture" in the sense that their own logic allows connecting ideas and establishing connections in a whimsical and haphazard way.

Apophenia (perceiving a connection or a significant pattern between unrelated objects or ideas) can be produced in altered reality games, in network role-playing games, or live action role playing. Who has not seen shapes, figures, or constellations on our bedroom walls at night or closet shadow-monsters during the summer naps of our childhood? Apophenia is a problem for designers and players of ARGs and related genres because it takes them away from the plot. But for QAnon this is not a bug, it is a feature. Q texts were the clues QAnons needed to reach their pre-established conclusions (Berkowitz 2020).

Berkowitz calls this "guided apophenia"—the established connections lead to the desired conclusions that have been created by the interpreters of Q.

> In clouds we can find all kinds of shapes, similarities, and every flicker is likely to be seen as a source of Morse code. The more information there is, the easier it is to allow apophenia to guide us to anything. It's about looking up at the sky and having someone point out constellations.

This false sense of discovery and the emotions it produces help overcome resistance (of friends and relatives) to the beliefs of QAnons. According to Berkowitz (2020): "guiding people to their conclusions is a perfect way to get people to accept a new and conflicting ideology."

Emotionally, QAnon is fueled by ancient hatreds—of Jews, Blacks, immigrant's women, outsiders, and intellectual elites. Thus it is easily spread through an increasingly polarized, fragile, and individualized social fabric. Berkowitz concludes that the very logics of discovery, of inquiry, together with the clues-as-riddles that sustain attention, engage high levels of emotional activation. This produced the energy to create a new community focused on bringing Trump back to power.

Therapists who specialize in violent cults say they are always endemic but can have virulent outbreaks, epidemics. Cults have developed effective ways of breaking down a potential host's resistance: isolation, love bombing, sex, drugs, sleep deprivation, massive consumption of sugar and complex carbohydrates, and repetitive intensive binging on the same content over and over. Sounds like a COVID-19 lockdown!

In the liminal zones between MAGA fanatics, Q's, anti-vaxers, New Age theocrats, racists, antisemites, and misogynist incels there are con-spirituality believers, who assert that Trump is a "light worker," aka a "crystal baby," an "Earth angel" or "star seed" (Karlis 2021a). Strangely, fantasies of pedophile rings and ritual cannibalism are much more widespread on the right than this "pastel Q" variant, as some term it.

One thing all the flavors of QAnon share is the grift. While "monetizing" content is common across almost all social media platforms, the way right-wing influencers extract money from their believers has reached new heights of hypocrisy and greed. Clare Birchall and Peter Knight (2023, p. 170) call it "Data Disinfo Capitalism" and they detail how profitable it is despite defunding attempts through deplatforming and payment apps refusing accounts. Such strategies come up against a difficult reality:

Social media platforms, search engines, data brokers and any other entities whose business models rely on the efficacy of data extraction infrastructure stand to gain the most in financial terms from the proliferation of COVID-19 conspiracy theories… (pp. 171–172)

As 2022 turned to 2023 first Twitter/X (at Elon Musk's direct command) and then Facebook and Instagram (facing stagnant growth) allowed Trump back on their platforms. After all, he is probably the single most profitable person ever to tweet or post for them.

David Icke, the man behind the lizard-people-rule-the-world story, brings in over $23 million a year (p. 172) but he is only one of thousands of "conspiracy entrepreneurs" that drive right-wing politics around the world into the crazy. They combine charging for content, selling their followers' attention to advertisers, and selling branded products to create multiple revenue streams. Their product (fear, anger, exceptionalism) packaged as principle, sells quite well. Birchall and Knight conclude, "Despite the vaunted idealism of anti-vax and anti-lockdown campaigns during the pandemic, they are often tied up with attempts to monetise their efforts. At the end of the day, it is nearly always about the grift" (p. 156).

Some rightists use Patreon, but most rely on payments from Adsense, YouTube, Amazon (which carries tens of thousands of QAnon and even Nazi products), Etsy (sells almost anything, same as Amazon), crowdfunding sites, and other players in the social media world. Some of the few sites that have been deplatformed, such as Red Pill run by Nemos, have established their own direct marketing systems (pp. 153–157). Without the profit, most of the famous voices of the extreme right would fall silent.

The spread of right-wing extremist ideology is a viral infection driven by exposure to multiple contagions through family, friends, and especially AI-powered for-profit social media. While reinforcement is required for transmission, only the susceptible become infected (Youngblood 2020). And we shouldn't forget that other ideas spread virally. One person's "extremist ideology" is another's necessary and glorious revolutionary movement. So how do specific ideas (BLM, white supremacy) catch on with specific people and how do we differentiate enlightenment from infection?

Why Black Lives Matter, Matters

Tell me how your community constructs its political sovereignty and I will tell you what forms your plagues will take.

—Paul Preciado (Bratton 2021, p. 2)

Fear. Panic. Loss of narrative control. You are the news now.

—Q (post #4348, May 29, 2020)

One of the things I love about living in downtown Santa Cruz—a block from City Hall, five from the town clock, a mile from the county building—is that many of the protests in our little town are nearby. But for most of 2020 and 2021 I didn't join them. During my life I have been to over a thousand protests in 20 different countries. I've helped organize several hundred of them: marches, rallies, dances, workshops, occupations, sit-ins, blockades etc. At many I was threatened by police, military, or right-wing activists, and a few times they tried to run me over. At some demos I was gassed, maced, and beaten by the police or even arrested. So why did I skip almost all protests in Santa Cruz and elsewhere for two years?

It wasn't because they weren't worth my energy. They were great actions to support—unions, including my own UC-AFT, to support Dreamers or BLM or the homeless. A number were aimed at the development plans of our little town's hipsterouise elite, now taking over from their parents, the hippieouise. When they are in power Santa Cruz has a great foreign policy and gives wonderful lip service to resisting climate change, racism, and misogyny, but locally they work for Seaside, the owner of the Boardwalk, and other corporations to foster hate and fear of the homeless, to subsidize giant hotels and parking garages as long as they aren't in wealthy neighborhoods, and enact NIMBY policies blocking trains and affordable housing.

But I just couldn't bear it. I didn't want to go out with dozens or even hundreds of people, even if they were wearing masks and many were my friends. I didn't just bunker into my little apartment, I bunkered into my mind. Somewhere in the middle of the pandemic I shook off some of the fear. My union was building for a strike, and even after I stepped away from teaching I helped with that. I was up at 5 am of the first day to go up and blockade UCSC again—I've been part of five shutdowns over the years—but at the last minute the UC Administration became reasonable when faced with running the university without lecturers, and we won a historic victory for adjunct faculty and other sectors of the precariat everywhere.

Each time I ventured out I had more energy for the next time. Being political is who I am. People have often asked me how I can stand to be so political, but it is not as if I have a choice. Ever since I was little I felt driven to pay attention and take sides. So during the pandemic I noticed a wide range of political reactions. Surprisingly enough, many anarchists, myself and the majority of my friends, were fine with following the latest science for wearing masks, avoiding people, testing oneself, and so on. Most anarchists believe scientists, not governments. If the government is listening to the scientists they seem the same, but they aren't.

The admirable CrimethInc Collective put out a "Surviving the Virus: An Anarchist Guide," that stressed Solidarity not Charity and advocated forming affinity groups (community bubbles) as part of larger and larger networks (2020, March 18). They called for a version of security culture that focuses on mutual aid, as the anarchist prince Peter Kropotkin termed it, "The fundamental idea here is that it is our bonds with others that keep us safe, not our protection from them or our power over them" (quoted in CrimethInc 2020).

Anarchists tend toward the philosophical. So it isn't surprising they realize,

The appearance of a new potentially lethal contagion compels all of us to think about how we relate to risk. What's worth risking our lives for? On reflection, most of us will conclude that—all other things being equal—risking our lives just to keep playing our role in capitalism is not worth it. On the other hand, it might be worth it to risk our lives to protect each other, to care for each other, to defend our freedom and the possibility of living in an egalitarian society.

Anarchists and others tracked how the pandemic was used as cover for more police, militarization, and direct attacks on refugee centers, anarchist bars, and even joggers in Sicily. Inevitably, anarchists started rent strikes, perhaps the first was Station 40 in San Francisco. Refusing to pay rent might seem like another utopian campaign, but the revenge of the real can reveal just how realistic anarchism can be. The range of the possible was shifted so much by the pandemic that radical demands eventually became moderate Democratic policies—rent relief, student debt freezes, and governments sending money directly to everyone, not just rich people!

Early in the pandemic the activist-artist Ian Alan Paul put forward "Ten Premises For a Pandemic" starting with: "A pandemic isn't a collection of viruses, but a social relation among people, mediated by viruses." He advocated solidarity and mutual aid, not reliance on the institutions that produced the crisis, and looked forward through the windows the pandemic opened. He saw, as many did during the lockdowns, that the life we were living in 2019 wasn't the only life possible. A different world is possible and it might be worse; a better world is possible, and it is necessary.

On the other side are people who only believe in the state and their own expertise. You can find as many of these on the left as the right. Benjamin Bratton's *The Revenge of the Real: Politics of a Post-pandemic World* (2021) is a particularly frustrating example. Half of his book is an appreciation of how real-world events collapse philosophical and political fantasies. The other half is proof that our own sacred beliefs are difficult to transcend. A Marxist philosopher who disdains every other approach, even other Marxist variants, much of his book is an attack on fellow leftists, including a wide range of laughable strawman assaults on anarchism, such as equating antifa with fascists. Marxists have long feared anarchism as particularly dangerous. It was official Marxist-Leninist doctrine in the Soviet Empire that the gullible revolutionary working class was easily fooled by anarchist ideals instead of following the leadership (Lenin, Stalin) of the dictatorship of the proletariat.

Bratton continues in this tradition, but what is more revealing is his effusive praise for the polices of the "Communist" Chinese government. When he wrote his book their zero policy seemed to have saved millions of lives but—revenge of the real—by late 2023 this was not the case. Those who think too much of expertise, especially their own, are dangerous. Whether it is medical experts who frame living science as completely settled, irrational conspiracy theorists/internet researchers who argue the testicular problems of a friend of a friend of their 2nd cousin proves something about vaccines, or Marxists who actually think their philosophy is an empirical science. You can expect them all to be often wrong.

The opposite approach to this top-down epistemology of the world is how social movements with diffuse leadership and perspectives, such as Black Lives Matter, understand the world. There are BLM nodes where spokespeople gather money and sometimes do great things and sometimes not. But the movement is countless groups and circles of friends and strangers coming together in outrage over the latest police murder. It isn't an ideology or a set of interlinked cells or pre-party formations aiming to be the vanguard or liberal pols planning how to get elected to Congress. It is many different people with many different views united by a sense of justice. This is why BLM protests when Native Americans, Whites, Asians, or Hispanic people are murdered by police, as they often are. All Lives Matter doesn't protest any of these, of course, because for them only Blue Lives Matter.

BLM would not exist without the digital phones that have attached themselves parasitically to most people. Again and again they have filmed police killing civilians. The police in the U.S. (and almost everywhere else) have always been violent and often behave illegally, but in the past it was what cops said against what victims said. No longer. Phone videos exposed so much police violence something like BLM became inevitable. Once it started, it spread through the Interweb and into the streets.

White people who have had friends and family murdered by the police, as I have, understood immediately that BLM was confronting both racism and police impunity, but most white people are blissfully unaware of what is really happening. It is a perfect example of what Steven W. Thrasher calls The Myth of White Immunity.

This myth tricks white people not only into making themselves needlessly susceptible to viruses, but also into refusing to see how at risk they are from all the harms and violence of society, which can grind one into the viral underclass. (2022, p. 231)

He goes on to note that while police in the U.S. kill people of color, in particular Native American and Black men, at extraordinarily high rates, overall more white people are killed by cops. All too often, the left seems more focused on how identity multiplies injustice than on injustice itself. If police murdered Black and Native American men at the per capita rate they kill white men, would that be just? Not for me. The current justice bureaucracy would not be acceptable even if it wasn't racist, even though it is an improvement over the whims of kings, emperors, popes, and caliphs.

Hugh Gusterson reminds us that justice was the "idealistic impulse underlying bureaucracy." The goal was "that everyone will be treated equally, fairly, and in accordance with rationally configured administrative procedures." It would be a "utopia of rules" (Gusterson 2019, p. 5). But actual bureaucracies have proven very corruptible since their origins, enforcing debt and taxes on the many for the few (Graeber 2011). But what if it wasn't people who made the bureaucratic decisions, what if it was algorithms? This is what the inventors, sellers, and proponents of algorithmic governance promise.

They have failed to deliver. It is people who write and manage algorithms, after all. Garbage in, garbage out; oppression in, oppression out. Algorithmic governance continues to spread dangerously without a real debate about its efficacy or ethics. Some governments and corporations pretend to care about AI ethics, but as we saw with Google, if the ethicists come up with the wrong answers they are purged (Vincent 2021). The pressures to create seductive applications to capture more and more profitable data and manipulate human behavior ever more effectively is increasing exponentially. AI and its growing power directly threaten democratic control of our complex techno-social human-machine civilization.

Controlling it starts with democratizing AI designs and applications by making their development transparent and participatory, and giving creators (coders, engineers, scientists) power over the process. Removing profit as the main motive for invention opens space for hacktivism, maktivism, and do-it-yourself/do-it-together movements. This is part of the larger campaign for democratic control of science and technology today.

How emotions and affect are exploited in algorithmic governance and AI is key. Emotions drive the majority of online behavior, and negative emotions are four times as powerful as positive. This is why social media algorithms send people to reactionary sites (Wong 2020). A vaccine skeptic is fed QAnon material because that is more likely to get them engaged. Reality be damned; there are profits involved, power as well. We need to go beyond the ethical questions surrounding this "race to the bottom of the brain stem" as it has been called. While "hacking of emotional attention" is a threat to democracy, it does produce seductive stories (Boler and Davis 2020).

We tell stories to account for the past, manage the present, and envision the future. For thousands of years culture was transmitted orally between generations; the arrival of the written word, literacy, is relatively recent. At each stage of human development new media can manifest extraordinary power to manipulate people. Radio for Hitler, Churchill, and FDR, television for Kennedy and Reagan, social media for AOC and Trump, are examples. But the dynamics of the process are much more complex, and only made more so by new digital technologies.

But even the latest techs are powered by the need to play. To be human is to play, even when it isn't fun we learn and change through play. But not necessarily for the better. Sara Grimes and Andrew Feenberg applied critical theory to the world of digital gaming by thinking through the role of play in culture and how reflexivity and boundedness "intertwine" with rules gives games their power (2012, p. 36). But power for what? QAnon shows games can catalyze politics taking control over people, their bodies, and reason. Yet games can also empower people. A diverse movement of coders and social entrepreneurs are pursing just that, referring to their work as serious games, not games, games for change, new gaming, persuasive games, and critical play. They build on a long history of games, play, celebration, and carnival as resistance.

QAnon needs to be confronted in ways beyond traditional leftist organizing. As culture makes culturally transforming technologies at an accelerating rate we must not only find new technologies that foster change, we need new understandings of change, such as speaking Virus. The transformative powers of technology and storytelling need to be harnessed: games to remake games, conspiracies to change conspiracies, stories not of terror and loathing (of self?), but of solidarity, sustainability, justice, and love.

6

Mutational Economies

Sacred Hunger

Money is sacred as everyone knows...So then must be the hunger for it and the means we use to obtain it. Once a man is in debt he becomes a flesh and blood form of money, a walking investment. You can do what you like with him, you can work him to death or you can sell him. This cannot be called cruelty or greed because we are seeking only to recover our investment and that is a sacred duty.

—*A slaver in Barry Unsworth's* Sacred Hunger *(1992, p. 176)*

Not so long ago, disasters were periods of social leveling, rare moments when atomized communities put divisions aside and pulled together. Increasingly, however, disasters are the opposite: they provide windows into a cruel and ruthlessly divided future in which money and race buy survival.

—*Naomi Klein (2008, p. 522)*

At 5:04 pm, PST, October 17, 1989, a powerful earthquake hit Santa Cruz. I was up at UCSC's Kresge College in the Italianate plaza. I remember being surprised when it just kept shaking and shaking. The large windows above me rippled like water. There was only one serious injury at the University, a friend of mine broke her arm jumping out a window. Down in town it wasn't so benign. Shawn McCormick, the barista at Santa Cruz Coffee Roasting Company who sold me the double latte I was still savoring a half hour later up at the U, was killed by a collapsing brick wall along with his co-worker Robin Ortiz. I've never been comfortable around brickwork since. Catherine Trieman lost her life when Ford's Department Store went down. Sixty-three died around the two bays of Monterey and San Francisco, almost 4,000 were seriously injured, and downtown Santa Cruz was savaged. The quake was centered just a few miles from our cabin in the Santa Cruz mountains. My ex, in our kitchen, just managed to save her toddler nephew from being brained by falling pots and pans. My son C, five then, was in a car in front of the house just coming home from a playdate. He thought the car had flat tires.

It took me a few hours to drive home instead of the usual 30 minutes, removing obstacles from Highway 9 with other motorists and people who lived nearby. A few places I had to take our little station wagon off the road to get past large trees or downed power lines. I wasn't home long before our local volunteer fire chief, with very different politics than mine to put it mildly, stopped by and asked me to join him in going door-to-door encouraging people in our neighborhood to evacuate because the Loch Lomond Dam up stream was considered suspect. He figured most of our often ornery and politically extreme mountain neighbors would believe at least one of us. So we evacuated down off the mountain and spent a few shaky days at the home of friends. The hundreds of little earthquakes that followed were worse on the nerves than the big one. By the time we moved back home, our only next door neighbor had already taken her family back to Texas. She couldn't stand the aftershocks.

Disasters on the news seem to happen quickly then trail off into funerals, restorations, and rebuilding. True enough, but living through a disaster is a lot slower. We were displaced for a few days, without water for a week, without power for several weeks, and without television for three months—something I heard about every day from my son. Our house was post and pier and rode out the quake just fine. We had an iron barrel stove with a pipe for heat, but there wasn't a brick chimney standing in the mountains as far as I could tell. Downtown Santa Cruz took years to recover and four decades to fill the last gap in Pacific Avenue.

COVID-19 killed over 300 in Santa Cruz, including my friend Will. Thousands more have Long COVID, many I know. Worse than the pandemic itself, is how society responded. Some things were good: the amazing vaccine production, the heroism of "essential workers," the realization that many essential workers were the worst paid—one of many viral underclass ironies. But other responses are deeply troubling when one looks toward the next viral disaster.

Irrational fears were used to mobilize political authoritarianism from the populist left (South Africa, Mexico, Nicaragua) to the extreme right.

Somehow, the many restrictions and crisis funding initiatives served to move even more of the world's wealth to the elite. As a report from the Institute of Policy Studies shows, at least $4 trillion in new wealth was captured by the world's 2,365 billionaires in just the first year of the pandemic, even as the world economy contracted 3.5% (Collins and Ocampo 2021).

Disaster capitalism is profiting from mass suffering (Klein 2008). But wasn't that the model of capitalism from its beginnings? It was birthed from European colonialism, which killed millions of people and extinguished several thousand cultures and their languages. Great swaths of nature were destroyed, eradicating hundreds of thousands of ecological niches driving countless species to extinction. A slavery/colonial economy was created that dominated the world for hundreds of years. Many countries remain trapped in the economic webs of their former colonizers, and oppressed by authoritarians who have adapted colonial systems to their own needs.

The impact of the pandemic on these countries was extreme, since almost everyone is in the viral underclass. In a report for the Arab NGO Network for Development, Gihan Abouzeid detailed the cost of COVID-19 in North Africa and the Middle East. Coming on top of the incredible inequality and underdevelopment in most of these countries, it was severe. Abouzeid notes that, "The pandemic affected all aspects of life, destroying the livelihoods of the poor and slowing down humanitarian support. The lockdown killed any glimpse of hope left, and depression spread almost as fast as the disease..." Falling international commerce and aid directly impacted a majority of people. When combined with the strain on already fragile health care systems weakened by years of "privatization contracts and austerity measures" pushed by foreign banks and other international institutions, it is no surprise at all that "many lives that could have been saved were lost" (2021, p. 3).

As with the viral underclass in developed countries, the impact of SARS-CoV-2 on what is a "deeply unequal world" aggravated the inequality. After all, the poor are "the least capable of self-isolation and protection." Tens of millions of ordinary people losing their jobs led to "alarming levels of hunger and suffering." Inevitably, female workers, who make up "the majority of healthcare workforce" were impacted the most. Even as wealth increases globally, the chasm between the rich and poor countries grows, and the gulf between rich and poor expands in those countries as well, leading to much higher "premature mortality" rates for workers and unemployed than the middle and upper classes (Abouzeid 2021, p. 7).

In Western Europe the top 10% of people control almost 40% of the wealth while the bottom 50% only 20%. In the U.S. the top 10% has 50% and the lowest fifth under 15%. There are wider gaps in Asia and most of Africa, and in the Middle East the top 10% controls over 60% of the wealth, and the bottom 50%, under 10%. Only South Africa is worse (Abouzeid 2021, p. 9). This is a clear result of colonialism and the current neo-colonial world order. The difficulty getting vaccines and administering them, using security services to preserve power while helping delivering medical and food aid, and other aspects of the pandemic in the Middle East can be traced to the same sources. Most of the West's interest in freedom for nonwestern countries is the freedom to economically exploit them. For the other empires, Russia and China, there is no pretense of freedom, just imperial elites negotiating with local despots in most cases.

Abouzeid does see some opportunities in the crisis. She supports UN Secretary General Antonio Guterres' call "for an immediate global ceasefire in all corners of the world." In terms of the Arab governments, she comments that, "This may be the first time in the history of Arab countries that they face the threat of a common enemy—a global pandemic—that does not come from a State or an army." It is a real chance for them to work together, and to gain legitimacy with their people by managing this crisis at least "somewhat successfully" (2021, p. 44). But most governments around the world, rich and poor, failed to do this. Only the corporations have been nimble, and that has been in pursuit of profits, not for the common good.

Driven by the Shock Doctrine, every disaster becomes a way for a greater proportion of the world's wealth to be moved to a tiny percentage of the world's people. This is accomplished by funding and controlling politicians, lobbying for the regulations that benefit the rich, capturing rebuilding and relief funds and shaping efforts to remake the infrastructure and the economy to their interests—gentrifying affordable neighborhoods, privatizing schools, subsidizing their projects and pollutions, society taking on their risks, cutting their taxes for "stimulus," and just giving them money. So it should be no surprise the COVID pandemic was no different.

From the start Naomi Klein warned that "big tech plans to profit from the pandemic" (2020). She called it a "Pandemic Shock Doctrine," a "Screen New Deal" version for high tech. Her prime example was the not-yet-disgraced NY Gov. Cuomo's push for a government funded private-profit surveillance capitalism infrastructure project: "Under Cover of Mass Death, Andrew Cuomo Calls in the Billionaires to Build a High-Tech Dystopia." Cuomo set up a "blue-ribbon" commission to reimagine post-Covid New York, led by former Google CEO Eric Schmidt. Big surprise, it was to be profitable tech integrated into everyone's life, whether they wanted it or not, linked to the Gates funded "smarter" (as in virtual) education system. Klein quotes Anuja Sonalker, CEO of a company selling self-parking technology: "There has been a distinct warming up to human-less, contactless technology…Humans are biohazards, machines are not."

Schmidt's other post-Google gigs were chairing the DoD's Defense Innovation Board and the National Security Commission on Artificial Intelligence. Both military AI and the "tele" economy depend on massive amounts of reiterative data processing. In 2019 Schmidt was pitching massive increases in military AI funding because China's programs were a threat to U.S. military dominance, advocating for the military-industrial complex President Eisenhower warned against in his Farewell Address.

At the heart of our economy the gargantuan military budget pumps away with no-bid contracts awarded by generals and colonels to the corporations they will soon go to work for, supervised by civilian officials appointed from those same companies who make sure the spending goes to the right districts where the right donations go to the right politicians. This is how banking and drug companies and the stock market and the carbon sector work as well. It is labeled disaster or surveillance capitalism in this chapter, and other types of capitalism in other places, but it is good to remember that it is "capitalist" not because it is the workings of a Free Market that can never exist but because those with the capital (wealth) makes the rules.

Over the door of the House of Siricus in Pompeii is still written *Salve Lucrum*: "Long Live Profit." Obviously, greed was not invented by Ayn Rand or even Adam Smith—who did *not* support a "free" market, by the way, believing a state was needed to keep the avarice of the powerful in check. Contemporary social critics who only want to blame the horrible state of the world on the relatively recent fantasy of Capitalism need to look deeper. It is today's incredibly powerful technologies driven by greed and corruption validating various ideologies, not just Capitalism, that is destroying the world.

To be fair, the system delivers. Maybe not an end to war or climate disaster or plastic pollution or poverty or the Sixth Great Extinction. But it does deliver unprecedented health, amusement, and comfort to not just the powerful, but billions of people from around the world. There may not be a future, but today there is sugar and fat, sex and religion, social and mass media, antibiotics, and vaccines for billions of us. But for just as many there is grinding poverty, horrific suffering, and unnecessary death.

When our body craves something—dark chocolate with pink Himalayan salt, tofu fried in bacon fat, perhaps a snap pea fresh off the vine, or just a long drink of clear, cool water—we seldom consider that hunger sacred. Perhaps the food is, blessed by the Divine or as a reflection of the holy nature we are all part of. But the hunger is biological, and it will end in energy, shit, muscle, and fat. Yet, when our heart or mind hungers we often assume our love (or just desire) for others, our belief in our Goddesses or Gods, the need to validate our working existence, is sacred. Ours is a narcissistic economy of desire.

The insanity of justifying slavery based on this sick logic drives Barry Unsworth's great novel *Sacred Hunger*. But is our world any less insane? Kleptocrats such as Putin and cult-of-personality dictators, cue Kim Jong-un, don't really need to justify their actions, but in democracies domination is a bit more difficult. Claiming any innovation that is profitable no matter how "disruptive," is not just justified, but better for everyone and therefore is morally good, is the myth that makes Capitalism so particularly dangerous. This desire to do well for oneself (which means good for all, doesn't it?) is the "sacred hunger" hidden behind the invisible hand, trickle down, the rise of all boats, the free market=democracy, fiduciary responsibility and all the other zombie lies that just won't die. This world is run by people who think the looting of living nature and human culture is not just inevitable but laudable, a holy act.

Slavoj Zizek argues that capitalism is not a virus. He is wrong. He admits it is "a parasite on us humans" and a "blind mechanism bent on expanded self-reproduction with total indifference to our suffering." But he says as it is not a biological virus it "doesn't exist in reality independently of us" depending as it does that "humans participate in the capitalist process." So, he announces, "capital is a spectral entity," echoing many virus=monster (ghost, parasite, zombie) tropes, as well as the "specter haunting Europe" *The Communist Manifesto* warned us about. He also admits that there are links "between the different levels of viral entities: biological viruses, digital viruses, capital as a viral entity." But not speaking Virus he doesn't see how to use this pandemic to go forward to a better world. Instead, he approves of Bruno Latour's call to make this a dress rehearsal for the Climate Crisis, for those who yet don't know that humans are part of nature (Zizek 2020, pp. 110-111).

But they are the same crisis; the viral polycrisis of exponentials climbing fast.

Surveillance Capitalism

Capitalism is the extraordinary belief that the nastiest of men for the nastiest of motives will somehow work for the benefit of all.

—*Often attributed to John Maynard Keynes*

Surveillance capitalism operates through unprecedented asymmetries in knowledge and the power that accrues to knowledge. Surveillance capitalists know everything about us, whereas their operations are designed to be unknowable to us. They accumulate vast domains of new knowledge from us, but not for us. They predict our futures for the sake of other's gain, not ours.

—*Johanna Zuboff (2019, p.11)*

My last car was infected. I called it the Sirius Virus. It was a 2009 white Taurus. Not the car I would have chosen. It was my parents' last car, and when they died I was the only son who needed a vehicle. Once I had it I noticed whenever I didn't drive for a few days—which is often because I've organized my life to mainly walk for shopping and entertainment—the battery would die. I got a new battery but that didn't help. My excellent mechanic tried to figure it out but no joy. Finally, after a year, I honed in on a strange detail: every time I'd get the car restarted after one of these battery failures, the radio would go crazy. Even if it was carefully turned off, it would leap to life and start screaming about how it couldn't find the Sirius radio satellite.

Some research on the Interweb revealed that I wasn't alone. It turns out that even if you didn't order it, and my parents did not, many cars come with the Sirius radio option. For some of them it has a hunk of code, a sick annoying algorithm, that automatically has it searching for that sweet Sirius satellite hookup even when the radio is off. Even when the car is off. And it cannot be fixed. Many have tried; all have failed. The car makers, and Sirius, are not interested in this problem. It seems to have been eliminated in new cars around ten years ago. That it haunts many older cars in use today is just bad luck for the owners.

So I put in a battery interrupt, and when I was not going to use the car in the next hour or two I opened the hood and disconnected the battery so the Sirius Virus could not do its thing. Then when I needed the car again, I'd open the hood, turn a nob, and reconnect the battery. Then the radio would go crazy with anguish, I'd turn it off and back on, turn down the volume which has automatically gone to high distress levels, find some blues or rock or reggaeton, and off I'd go. I learned to live with my car's incurable chronic infection. Sometimes everything seems infected.

If you use apps on your smart phone you are probably being monitored by over 5,000 e-trackers per week. Yes, iPhones too. The social media leviathans are tracking your computers as well and your accounts in the cloud. They capture the data that they use to sell us music, politicians, books, games, and so much more. No wonder Shoshana Zuboff's *Surveillance Capitalism* is so scary. She has analyzed the dominant companies of our age (Google, Facebook/Meta, Amazon, Apple, Microsoft/LinkedIn, Verizon/Oath) and argues convincingly that they are producing a new economic form—surveillance capitalism.

Zuboff thinks this surveillance capitalism is "rogue" because it replaces "the continuous intensification" of production by industrial capitalism with "the continuous intensification of behavioral modification." But it is hard to see that today's dominant industries are any more effectively evil, or less, than the Robber Barons. The technologies have changed, people have organized, so new forms of control are needed. Zuboff argues Google "invented and perfected" this new type of capitalism, just as General Motors did "managerial capitalism" (2019, p. 9). With shocking admissions from key corporate players, she reveals that this new system is a bit more complicated than the cute, "if it is free, you are the product." The product isn't us, it's data about our behavior, which leads to predictions about what we might want and do in the future and, more and more, ways to shape who we are. Surveillance capitalism's actual customers are the enterprises that trade in its markets for future behavior (Zuboff 2019, p. 10). This has been made possible by the development of Algorithmic Intelligence.

Often critics blame the hardware and software, the tech, for the way the world is. Yes, new techs come out of old political economies and have designed-in affordances for profit and hierarchy, but the past is not definitive. In the end it isn't the tech, it is what is done with it.

> Surveillance capitalism is a market form that is unimaginable outside the digital milieu, but it is not the same as 'digital'...the digital can take many forms depending upon the social and economic logics that bring it to life. It is capitalism that assigns the price tag of subjugation and helplessness, not the technology. (Zuboff 2019, p. 15)

Zuboff calls the social media companies Big Other, and argues they have developed a new form of instrumentalist power. This signals the transformation of the market into a project of total certainty, an undertaking that is unimaginable outside the digital milieu and the logic of surveillance capitalism (Zuboff 2019, p. 20). It starts with monopolizing knowledge. Instead of the division of labor that undergirds industrial capitalism, there is a new division of learning. A restricting of knowledge, authority and power that Zuboff (2019, p. 181) maps into three fundamental questions:

1) Who knows? Who gets to learn what?
2) Who decides? Who decides who gets to learn and what?
3) Who decides who decides? Who answers the first two questions?

The learning here isn't the formal education we all go through. That is important, but it is a well-known process with deep historical roots. It is more about all the new knowledge our machines now collect and analyze. It is big data. Big Other needs unfettered access to all it can collect in order to produce surplus value, for profits and the power to make more profits. This surplus is the knowledge of human behavior from the monitoring of our emotional lives, the lived libidinal economy. Thus,

Surveillance capitalism departs from the history of market capitalism in three startling ways. First, it insists on the privilege of unfettered freedom *and* knowledge. Second, it abandons long-standing organic reciprocities with people. Third, the specter of life in the hive betrays a collectivist societal vision sustained by radical indifference and its material expression in Big Other. (Zuboff 2019, p. 496)

This is only possible through massive repetitive digital processing. Material technology is a precondition but cultural forces (the chicken before the egg) are the great shapers of political-economies. Culture is not superstructure, it drives the forms economies take. Law constrains, corruption facilitates, and emotions makes some algorithms pernicious and profitable and others not. The relentless quest for more profits has created the world we live in. The world could be different, better, with the basic technologies we have. Nowhere is this clearer than the digital realm, but it is just as real on the ground, the extracted lands, the warming carbonated air, the living systems choking on micro plastics while they die in the heat.

Sadi Plant and Nick Land say in "Cyberpositive" that "Capitalism is not a human invention, but a viral contagion, replicated cybernetically across post-human space [...] if schizophrenia is not yet virally programmed it will be in the future" (Plant and Land 1994, paras 14, 19). But Capitalism isn't the only problem. After all, it isn't just surveillance capitalism, we also have surveillance autocracy (China) and surveillance kleptocracy (Russia), and everywhere corruption is a problem, multiplying whatever injustices are "normal" for a particular society.

Corruption is a conspiracy and it is not limited to nation-states. Many international organizations are choking on it, such as the UN and its public health arm WHO. Because WHO depends on donors for 75% of its budget (as of 2015), WHO's policies have long reflected their interests. In 2004–2005, 91% of donations went to diseases that were only 8% of yearly human mortality. The dues from countries, often late or even withheld, are spent at closed meetings dominated by corporate interests. So Malaria spending goes to insecticide makers (who want the danger to continue) and WHO doesn't buy generic drugs because of the power of pharmaceutical companies. WHO's failures combating Ebola and other outbreaks can be traced to this, and on the sad reality that local WHO functionaries are appointed by local governments that are often in denial and/or deeply corrupt (Shah 2016, pp. 117–119).

We must confront corruption. Corruption is corrosive to democracy, and it strangles civil society in petty dictatorships. When I visited Egypt soon after the overthrow of Mubarak, three different people told me they knew the revolution was real when they went to the Department of Motor Vehicles and did not have to pay a bribe to get service. But what if the very form of governance is legal but unfair? What if Veillance Society is run for profit?

Going Full Circle

[W]e are all committing suicide like a stupid virus that kills too many of its hosts: the planet and society. What is the kernel of that virus of capitalism, the virus of perfect maximization? I think it has to do with shareholding. The one market that is the beginning of all this toxicity is the share market. We have come to accept the idea of the corporation as a given. Lots of people work in it, but none of them own it. The people who own it? We don't even know who they are. There are corporations inside other corporations and shell corporations and so on.

—Yanis Varoufakis (Waters and Varoufakis 2020, p. 254)

In David Eggers' (2013) powerful and disturbing novel *The Circle*, a young Customer Experience Tech named Mae joins a company called The Circle. It dominates the social media and informational economies, combining the power of Alphabet/Google, Apple, Amazon, and Facebook/Meta through viral algorithms. The entrepreneurs of The Circle claim that their intimate and pervasive collection of all possible personal information is for the public good. No doubt it is an accident that following their principles—*Secrets Are Lies*, *Sharing Is Caring*, and *Privacy Is Theft*—makes them incredibly rich and powerful. They take over voting, crowd-source a hyper-panoptic culture to enforce everything from recycling to an end of privacy, and end up dominating (one might say even absorbing) most of the economic and social life of the world. Mae not only comes to accept The Circle but becomes one of its leading advocates and joins upper management.

For several years we required *The Circle* as summer reading for frosh at my STEM focused college at the University of California at Santa Cruz. A third of my students thought it was a utopia, a third thought it dystopian, a third were very confused indeed. The same rough divisions are to be found in their attitudes toward the social media behemoths *The Circle* warns us about.

In the real world, Shoshanna Zuboff admonishes us that,

> It is important to understand that surveillance capitalists are impelled to pursue lawlessness by the logic of their own creation. Google and Facebook vigorously lobby to kill online privacy protection, limit regulation, weaken or block privacy-enhancing legislation, and thwart every attempt to circumscribe their practices because such laws are existential threats to the frictionless flow of behavioral surplus. (2019, p. 105)

Clearly, The Circle's corporate principles are alive and metastasizing. I already liked *The Circle* but after reading *The Age of Surveillance Capitalism* I now consider it is as important a warning as *Brave New World*. If civilization makes it through the next 50 years, my great fear is it will be a combination of *Brave New World* and *The Circle*. *1984* seems unlikely. Orwell's nightmare is far removed from what students imagine is possible. But turning the source of their greatest pleasures into dystopia, that shocks them. It scares me.

One of the dissidents to The Circle's regime of surveillance capitalism is Mae's oldest friend, Mercer. Before he is hounded to death by Mae and a Circle-catalyzed mob demanding access to his life, he explains to Mae why he doesn't want to join.

> I'm social enough. But the tools you guys create actually *manufacture* unnaturally extreme social needs. No one needs the level of contact you're purveying. It improves nothing. It's not nourishing. It's like snack food. You know how they engineer this food? They scientifically determine precisely how much salt and fat they need to include to keep you eating. You're not hungry, you don't need the food, it does nothing for you, but you keep eating these empty calories. This is what you're pushing. Same thing. Endless empty calories, but the digital-social equivalent. And you calibrate it so it's equally addictive. (Eggers 2013, pp. 133-134)

As with gambling machines, fast food, mass media, electronic games and many other aspects of postmodern life, social media are designed to be addictive (Schüll 2014). So it is no surprise that by the end of the book Mae has got her best friend killed, alienated her parents by bugging their house and exposing their sex life to the world, and betrayed The Circle's one ethical founder, who was trying to kill the monster he made by revealing that the other two founders, Stenton and Bailey, were not monitored themselves and were planning on taking over the world...for everyone's good, of course. On the last page (Eggers 2013, p. 491), she is at the bedside of her unconscious friend Annie, the woman who got her the job at The Circle, who has had a breakdown. Mae is aggrieved.

Another burst of color appeared on the screen monitoring the workings of Annie's mind. Mae reached out to touch her forehead, marveling at the distance this flesh put between them. What was going on in that head of hers? It was exasperating, really, Mae thought, not knowing. It was an affront, a deprivation, to herself and to the world. She would bring this up with Stenton and Bailey, with the Gang of 40, at the earliest opportunity. They need to talk about Annie, the thoughts she was thinking. Why shouldn't they know them? The world deserved nothing less and would not wait. (Eggers 2013, p. 491)

The film turns this ending on its head and Mae (Emma Watson staying true to her Hermione character) exposes Stenton and Bailey for the evil, greedy, hypocrites they are. But that's Hollywood, not literature (Reilly 2017). In the real world, as in the book, things are a great deal worse. For as Zuboff documents, the next step of surveillance capitalism is the quantification of the physical world and our private mental worlds will follow. Considering the progress being made with "mind reading" neuroscience research using many of the same viral AI digital technologies as surveillance capitalism (machine learning, big data on human mentation/behavior, big data mining and manipulation) it probably will be soon (Gray 2014a).

The canary in the coal mine is the fat little Pokémon. Collecting and using real world data is the inevitable next step. It started with Google Maps. Zuboff quotes Brian McClendon, Goggle Map's senior product manager who said in 2012,

If you look at the offline world, the real world in which we live, that information is not entirely online. Increasingly as we go about our lives, we are trying to bridge that gap between what we see in the real world and [the online world], and Maps plays that part. (2019, p. 150-151)

Why is Google so anxious to bridge that gap, not just with Google Maps but also Street View, and even giving backpack cameras to tourist boards and hiking clubs to capture images that were once "off the grid" (Zuboff 2019, p. 152)? Because the real world is data that can be leveraged for much more power. Zuboff explains, this is about...

...the migration from an online data source to a real-world monitor to an advisor to an active shepherd—from knowledge to influence to control. Ultimately, Street View's elaborate data would become the basis for another complex of spectacular Google incursions: the self-driving car and "Google City"...These programs aim to take surplus capture to new levels while opening up substantial new frontiers for the establishment of behavior futures markets in the real world of goods and services. It is important to understand that each level of innovation builds on the one before and that all are united in one aim, the extraction of behavior surplus at scale. (2019, p. 153)

Add in phone data and all you can get from wearables, such as the progeny of Google Glass, and AR games like Pokémon and the Harry Potter spin-off Wizards Unite, and you are sucking up an incredible amount of real world/human behavior data. This gives the old truism "Maps Make Empires" a whole new meaning. This is why all of these companies invest massively in viral machine learning because it depends on billions of copies of simple algorithms that mutate as they are exposed to more input. It isn't really intelligence in any human sense that they are creating, it is the ability to mimic intelligence enough to interact with humans (Hi Alexa! Hi Siri!).

Google Maps fits into an analogy Zuboff uses throughout her text: colonization. Mapping was integral to European colonialism. Mapping is part of conquest because for conquest to be worthwhile there must be control of what (whom) is captured. As an anonymous Internet of Things designer told her:

We are learning now to write the music, and then we let the music make them dance. We can engineer the context around a particular behavior and force change that way. Context-aware data allows us to tie together your emotions, your cognitive functions, your vital signs, etcetera. We can know if you shouldn't be driving, and we can just shut your car down. We can tell the fridge, "Hey, lock up because he shouldn't be eating," or we tell the TV to shut off and make you get some sleep, or the chair to start shaking because you shouldn't be sitting so long, or the faucet to turn on because you need to drink more water. (2019, pp. 295–296)

See, nothing to worry about. Having your appliances boss you around is just for your own good and the profit of Big Other, of course. Drawing from operant conditioning and their own research, these helpful engineers have developed a whole suite of operationalized manipulations in the form of viral algorithms for "conditioning at scale." There are at least 93 such "tools" identified in an important analysis from 2014 (Lyons et al., in Zuboff 2019, p. 621, n. 2) including: scheduled consequences, reward and threat, repetition and substitution, antecedents, associations, feedback and monitoring, goals and planning, social support, comparison of behavior, communication of natural consequences, self-belief, comparison of outcomes, shaping knowledge, regulation, identity, and overt learning.

It doesn't take much to get them to admit their goal is control. As "the chief data scientist for a much-admired Silicon Valley Education Company" told Zuboff, anonymously:

The goal of everything we do is to change people's actual behavior at scale. We want to figure out the construction of changing a person's behavior, and then we want to change how lots of people are making their day-to-day decisions. When people use our app, we can capture their behaviors and identify good and bad. Then we develop 'treatments' or 'data pellets' that select good behaviors. We can test how actionable our cues are for them, and how profitable certain behaviors are for us. (2019, p. 297)

Where do they get the unmitigated gall to work toward this level of interference in people's actual lives? They use our lives as raw data which they consume and shit out as "data pellets" to manipulate us. What is their justification? It is profitable. If it makes money they have every right to do it. Money is sacred. Their hunger for it is sacred.

So far, this exploitation of the many by the few has proven distressingly nimble. In this insane system the profit goes to the rich who also get the credit for the work of the many that has led to the today's world: improved health, a decline in violence and poverty, and technological marvels that benefit billions of people. This progress isn't because of systems of exploitation but in spite of capitalism and other forms of brutal profit-taking. The original sin of avarice is baked into the affordances of technologies by avarice.

But if divorced from greed these same techs can empower citizens, enable liberatory hacking, and disrupt not just markets, but reified political systems and their elitist masters. Tech makes their surveillance more powerful, and yet it makes resistance stronger—if used differently, not for profit but for sustainability, not for domination but justice, not out of anger and fear about our nature (and thus all nature) but rather—is there a better word?—love.

It can be innovative as well, and propagate exponentially.

Spreading Innovations and Digital Proliferations

These similarities are striking. An external event strikes. Fear grips the system, which, in consequences seizes. The resulting collateral damage is wide and deep...You couldn't tell a story about why Lehman had brought the financial system down without telling a contagion story.

—Amy Haldane, Chief Economist at the Bank of England comparing the SARS epidemic to the 2008 financial crisis. (Quoted in Kucharski, pp. 72, 74)

That's why we need to be debating democratic models of effective response to present and future plagues, ones that mobilize popular courage, put science in command and use the resources of a comprehensive system of universal health coverage and public medicine. Otherwise we cede leadership in the age of constant emergency to our tyrants.

—Mike Davis (2020, p. 44)

In the middle of the pandemic I stopped teaching. It wasn't because it was all online now, although that didn't help. In the decades I have been teaching I have often done it virtually. While a professor in Montana from 1996 to 2005 I always had some distance courses. Many of my students lived out on ranches or on reservations like Rocky Boy and Blackfeet Nation. With our techs I made videos that we mailed to the students, and then we'd have class by telephone conference calls one or two times a week. Some students I never met until they graduated. Some I never met at all. I also taught at Goddard College and The Union Institute and University in the 1990s and early 2000s. Both used an intensive residency model, where once a semester we'd spend a week or so together planning their work and then they would send it to me. Starting in 2016 I taught distance courses for New York University's Tandon School of Engineering on NYU's own platform, with weekly modules with lectures and interactive elements, along with real-time meetings in virtual salons and for various presentations in Zoom.

So when my UCSC classes were Zoomed into virtuality I was much better prepared than most of my colleagues. But it was still hell. Teaching is a profoundly interactive and intimate experience. It is always uphill when virtual and even more emotionally draining than in-person classes.

It didn't help that my students were all first year, in the Core Course introducing university norms, crucial reading skills, and scholarship about their college's theme. At my home, Crown College, that is "The Ethical and Social Implications of Emerging Technology." At College Ten (now John L. Lewis College), I taught "Community and Social Justice."

Even using the best tech nothing beats embodied learning. For weaker students face-to-face is crucial, and for the better ones it is better. I try to avoid teaching because it takes all my psychological energy to do it well. I have little left for writing, which is my true love, my actual compulsion. Teaching is honorable work and I like it, but I don't love it. But sometimes I need the money.

I've had some jobs in my life that were considered menial but that the pandemic magically revealed to be essential, not that the compensation then or today reflects this. Other jobs, not so much. Looking back at what I've had to do to survive I'd say a third of the work was bad for me and for the world. Somebody made money though. The rest was either neutral, as in too hard to judge morally (handyman in the Stanford Physics Department) or even good for the world (teaching, child care, elder care, gardening, insulating houses). While all of us are in the economy much of the most important work isn't paid, at least with money, like child raising and home keeping. But some of the most damaging work for the world, most of finance for example, is the best compensated of all.

Consider the recent invention of correlation trading. Insurers noticed that people were much more likely to die after their spouse did ("broken-heart syndrome") and set insurance rates accordingly. The idea that correlations could predict mortgage payments became popular among financiers, and they went on to invent many "instruments" bundling them together. They claimed a bunch of bad mortgages together transmuted into good investments. Unfortunately for them, and the housing market in 2008, this was "far removed from reality," and the models driving them depended on a "mathematical illusion." As the credit specialist Janet Tavakoli noted a year before Bear Stearns collapsed.

Correlation trading has spread through the psyche of the financial markets like a highly infectious thought virus. So far, there have been few fatalities, but several victims have fallen ill, and the disease is rapidly spreading. (Kucharski 2020, pp. 42–43)

The ecologist George Sugihara mainly focuses on marine conservation, but he did work for Deutsche Bank for four years. He used his experience predicting fish stocks to model financial stocks. "Basically, I modeled the fear and greed of mobs that trade." An article in *The Lancet* noted the similarities as well. "The recent rise in financial assets and the subsequent crash have rather precisely the same shape as the typical rise and fall of cases in an outbreak of measles or other infection," wrote Robert May, another scientist who was seduced into financial modeling by the money. Sugihara was his student. May, a zoologist, added that, "When something is going up without a convincing explanation about why it's going up, that really is an illustration of the foolishness of the people" (Kucharski 2020, pp. 44–45).

Crypto currency is another viral financial mania that is also a dangerous infection. Nicholas Weaver, Senior staff researcher at the International Computer Science Institute is clear:

This is a virus. Its harms are substantial. It has enabled billion dollar criminal enterprises. It has enabled venture capitalists to do securities fraud as their business. It has sucked people in. So either avoid it or help me make it die in a fire. (Robinson 2022)

It also has horrible environmental effects. In 2022 it was consuming more energy than Egypt, more than 1/200th of all electricity production in the world. They invent made-up money to fuel a Ponzi scheme basing it on meaningless computing that has become a major contributor to the climate crisis that might destroy civilization. For profit.

No wonder Zuboff sees the danger as more existential even than "enslavement." Consider the details of the behavioral-libidinal economy of today's dominant economic organizations and the emergence of a surveillance economy.

The commodification of behavior under the conditions of surveillance capitalism pivots us toward a societal future in which an exclusive division of learning is protected by secrecy, indecipherability, and expertise. Even when knowledge derived from your behavior is fed back to you in the first text as a quid pro quo for participation, the parallel secret operations of the shadow text capture surplus for crafting into prediction products destined for other marketplaces that are *about you* rather than *for you*. These markets do not depend upon you except as a source of raw material from which surplus is derived, and then as a target for guaranteed outcomes. We have no formal control because we are not essential to the market action. In this future we are exiles from our own behavior, denied access to or control over knowledge derived from our experience. Knowledge, authority, and power rest with surveillance capital for which we are merely 'human natural resources.' (2019, p. 328)

But social media isn't just a form of exploitation; it also can be a tool of liberation. Idle No More was founded on Facebook by four First Nation women in Canada, and became a worldwide network of thousands in two months. It is not the only movement to use digital technology integrally. In many ways, labels such as Arab Spring, Occupy, Los Indignados and Idle No More don't mark new movements but rather incarnations of ongoing resistance cultures that social media revitalizes in new and more potent ways.

The spark that started the Arab Spring was the protest suicide of Mohamed Bouazizi, a Tunisian street vendor the police forced out of business for insufficient bribery. His sacrifice sparked a conflagration. Social media played an important role in shaping the understandings (the fuel) of the Tunisian revolutionaries, and it was even more important as revolution spread to Egypt, catalyzed by the police murder of the blogger Khaled Mohamed Saeed and beyond.

Early on, Anonymous launched "Operation Tunisia." It included Denial of Services attacks on government websites and a care packet in Arabic and French for cyber dissidents on identity concealment. A Greasemonkey script was written to "help Tunisians evade an extensive phishing campaign carried out by the government." Anonymous was not alone. Among the important cybergroups helping the Arab Spring were Avaaz, elecomix, Witness, Heridict, OpenNet Initiativ, and the Open Mesh Project.

Still, social media doesn't guarantee the success of liberation movements, nor that authoritarians can't use it as well. The sociologist Zeynep Tufekci explains in *Twitter and Tear Gas*,

Historian Melvin Kranzberg's famous dictum holds true: 'Technology is neither good nor bad; nor is it neutral.' Neither are technology's effects static; everything evolves as people invent, innovate, and appropriate technologies for their purposes. This dynamism does not mean that technology provides a level playing field, where each side is equally empowered and equally able to appropriate technologies for its purposes. (2017, p. 262)

We can chose to be good or we can chose to be bad, but we can't chose to be neutral.

Zuboff does look for hope in the realization that the target of capitalism is now human nature, having pretty much despoiled non-human nature. In the conquest of wild nature industrial capitalism's victims were mute. Those who would try to conquer human nature will find their intended victims full of voice, ready to name danger and defeat it (Zuboff 2019, p. 525). People did, and do, resist the destruction of nonhuman nature, but not enough. Years ago we were warned by Blake about the Satanic Mills. Klein has revealed disaster capitalism and Zuboff delineated surveillance capitalism and its ambitions. Mike Davis sounded one of the clearest alarms in his last book: *The Monster Enters: COVID-19, Avian Flu and the Plagues of Capitalism.* He noted that around the world governments used the pandemic to repress democracy, and in the case of Viktor Orban in Hungry implementing martial law in the world's first "coronavirus coup" and the first dictatorship in Europe since the death of Franco. But Davis saw reasons for hope as well.

Many writers and scientists, poets and carpenters, mothers, fathers, plumbers and ramblers, eco-activists and other rebels against extinction, have made saving nature their priority. There is a dialectic at work (although not Hegelian and even less Marxist) so that each seductive (dreamy *and* nightmarish) step toward destruction is met with growing resistance.

Whatever the affordances of the techs we have, I believe that we can win...all the evidence to the contrary. What's the option? Without hope agency is impossible. And there is Virus. Exponentials can run both ways as we see in the next chapter, which is the story of how cyberpunk has inevitably arisen to speak against the emerging digital dystopia.

7

Speaking Virus—Cyberpunk Style

The Dialect of Cyberpunk

"This Snow Crash thing—is it a virus, a drug, or a religion?"
Juanita shrugs. "What's the difference?"

—Hiro and Juanita debate metaphysics in Snow Crash by
Neal Stephenson (1992, p. 200)

People are a virus.

—Agent Smith in The Matrix

This world we now live in has been coming my whole life. I have never thought any time was normal. I know most people—not you, dear readers, who obviously have a richer perspective—have trouble accepting that the past, with its bizarre tragedies and beauties, is real; or how real the very different lives of people on the other side of the world are today, on the other side of town right now, the part they never visit. But for me it is impossible *not* to accept these are part of our reality—The Black Death, the Holocaust, climate disaster, the bloody wars of Syria, Iraq, Afghanistan, the Ukraine, and Gaza, and on and on. I own this. As long as I can remember, I've inhaled both history and the daily news. Perhaps it was coming of consciousness first in Vietnam. My continuous memory started on our two-day six-stop Pan Am flight to 1957 Saigon. My mother, my two brothers, and I were on our way to join my father, working as a highway engineer for a civilian contractor, building infrastructure, as it were. It was a quiet period in the long Indochina War, but still war was all around.

Or maybe it is because books are as much a part of real existence for me as the blood, sweat, and tears of embodied life. Or maybe I've just been a curmudgeon since I was six, as my brothers claim, or an "old soul" as more than one lover has complained, annoyed at my "long view" of this problem or that. I've lived all my life in an unstable world of continually expanding science, technology, pollution, and extinctions. A world where war is ubiquitous, revolutions are possible, genocides too, and pandemics aren't just likely, they are inevitable. In any event, when I first encountered the cynical turbo-warped technoscientific the-future-is-now saturated perspective of cyberpunk, with that little glow of hope off in a corner, I remember thinking, "Yah, this is how it is."

Credit exponential change. Humans are a young species that changes at an extraordinary rate biologically and culturally. Look at what we've done to the planet! Different frames of reference have helped me navigate a world which is relentlessly changing—Literature, History, Anarchism, Science, Feminism. I think of these as languages, each with their own grammar and vocabularies, with ways of thinking that illuminate what the daily discourse of mass media, of work, of our friends, does not. Of course, they aren't literally languages. You can think of them as dialects if you wish, or genres. They are conceptual but real. They overlap and interlink as languages do. The political traditions and ideals that frame my world view are close to each other in my mind, like Spanish and Catalan and Italian....but they are not the same. Locked down in 2020, I realized I needed to develop some new understandings, a language that zoomed in on the key dynamics that were making that year so horrible: COVID-19, fires, fascism.

I watched it, online, on TV, out my window. I read it in texts and the faces of loved ones and strangers. I thought about how we watched and were watched more and more through technology, as science fiction dreams and nightmares unfold before us. It was the cyberpunk world I'd inhabited since the 1980s, but more so, as cyberpunk has spawned biopunk, steampunk, solarpunk, and a dozen more. Most overlap with CliFi, climate fiction that ironically is about the truth of global warming. *Virus as a Language* is part of the same tree. The family resemblance, as Wittgenstein would put it, is unmistakable.

One of the original cyberpunks, Bruce Sterling, noted us boomers grew up not only reading science fiction, but actually living "in a truly science-fictional world." The "extrapolation" and "technological literacy" of SF aren't just "literary tools," they are "an aid to daily life...means of understanding" the world we live in (Sterling 1986, pp. x–xi). Or, in the words of a second generation cyberpunk, Paolo Bacigalupi, author of the *The Water Knife* among other near future scary-as-hell-the-world-is-dying CliFi novels, cyberpunk chronicles "the future that we are actively building," like it or not. It is urgent. It wasn't "the future 15 minutes out—it was the future sideswiping you and leaving you in a full-body cast as it passed by" (2012). Our age is relentlessly cyberpunk. But why *cyber...punk*?

Punk is what bullies call their victims. Punk is the resistance of the underdog. The word has a mysterious history, appearing in various forms in the 1600s. It apparently was Native American for rotted wood and tinder and in England became slang for a prostitute. Eventually, it turned into a self-descriptor of pride. Anti-authoritarians who were both anti-fashion and anti-rock/pop music claimed the label with a vengeance in the 1980s, rejecting the authority of church, state, corporations, and popular culture.

That it was almost immediately turned into a brand, with its own fashion rules and pricey boutiques in Soho, Koenji, the Bowery, and the Tenderloin, was more than ironic, it was inevitable. Less predictable, but more important, was that a number of SF writers soon added cyber and began labeling their near-future digitally saturated hacktivist-as-antihero stories cyberpunk.

Cyber is short for cybernetics. *Kyber* is ancient Greek for the feedback processes of complicated systems, such as the calculations someone steering a ship (*kubernetes*) has to do automatically. Aristotle applied it to the give and take, the feedback loops if you will, of politics. The great mathematician Norbert Weiner adopted it for the study of the way information flows and controls all systems, living or not.

Viruses of all types only exist within complex systems. They can't survive on their own; they are parasites. This is true of viruses in digital, cultural, and biological systems. In SF, their fundamental similarity is often made manifest. Consider the great novel *Snow Crash*. It is named after the "drug" Snow Crash, which started as aspects of the Sumerian language (culture) that could act on the brainstem of humans (biology). This linguistic virus was personified by the goddess Asherah and mainly spread through computer gamers (digital) to all humanity. In the story, the god Enki made an antidote, a counter-program which led to the Tower of Babel effect, the splintering of human language so that Snow Crash could not spread. In the story, some evil fucks are trying to bring it back through a drug taken to immerse oneself in games. It spreads culturally, digitally, and biologically as gamers recruit new victims to play. Sound familiar?

Real plagues can combine elements almost as surprising as the Snow Crash fantasy. In 2005 the Corrupted Blood plague "killed" hundreds of thousands of players in the World of Warcraft online role-playing game. It started as a weapon to use against a blood stealing monster called Hakkar, but bad programming allowed it to spread throughout the world, helped along by "griefers" who infected people's on-line personas, their avatars, on purpose. Only the most powerful avatars could survive it. Digital patches failed so the game had to undergo a hard reset, destroying all the avatars. For many people, whose avatars had been an actual part of their personality for years, it was a real disaster. The line between our biological and digital selves is often hard to find.

Back in the late 1980s I interviewed Manfred Clynes, who coined the term cyborg. We exchanged several drafts of the interview by email until, one day, I got a phone call from him. Without preamble he screamed, "You have infected me!" It took me some time to calm him down and get him to explain that his anti-viral software had detected a virus in a file I sent him—a very benign virus had that slipped past Oregon State's antiviral programs. But for Manfred it was real and scary.

Is our world that different? The spread of SARS-CoV-2 involves a similar interplay of viral dynamics. The Corrupted Blood plague and planned pandemics in Warcraft and other online games are now studied by epidemiologists, who appreciate the wealth of data available in tracking digital infections. Epidemiologists, statisticians who study the spread of extremism through the Interweb, anthropologists of mobs and cults and political movements, all have to think virally. Some genres of writing also have viral thinking as integral to their perspective, and foremost among these is cyberpunk. Cyberpunks have always spoken Virus. These never-utopian, digitally mediated punk-inflected visions of the near future feature viruses of all major typologies—cultural, biological, digital—as central organizing principles of society. We can see this in real time through the vast web of viewing technologies, from satellite feeds on-line to live casts from war zones to millions of posts (tweets on X), emails, and web articles competing 60-60-24-7-365 for our attention and trust.

You can feel the thrum of it, a soundless moan in your head and itch between your shoulder blades. It is the CCTV cameras on every urban corner, hulking like vultures; the torrents of metadata streaming by the millisecond to the NSA's Bumblehive, a million square feet of servers in Utah; drones buzzing and proliferating and shrinking; phone cameras ready to stream to the web every kitten or killing they see. Your searches are haunted by stupid ghosts sent by Amazon, Google, Apple, and Facebook to sell you shit you just bought; doxing, identity theft and COVID-19 are taking out your friends left and right; and that shadow of yourself, your medical data and buying habits and Pinterest page of old union posters, your porn predilections—your digital doppleganger—is growing in cyberspace every keystroke you make. Welcome to today. Welcome to infectious, sometimes fatal (to life and reason), exponentially spreading Veillance Society.

Veillance, from the French "observed," from above (*sur*), from below (*sous*), from almost every angle. We were warned of its coming by many, but perhaps most helpfully by the cyberpunks. Because their vision of a dangerous, darkening (for oh so many reasons), increasingly corporate world saturated in new neon bright digital do-dads emerging every hour from a festering Silicon Valley and its regional infections from Bangalore to Nairobi to Austin, while the line between flesh and machine dissolves in accelerating and ever more interesting ways, is clearly coming true. "Cyber" because reality is being digitized and "punk" because we know the process, driven as it is by greed (for fame, wealth, and power) is both corrupted/corrupting and essential for who we are. It is our worst nightmare and our greatest hope.

"Watch or be watched," Pretty Boy Credo declared in Pat Cardigan's" Pretty Boy Crossover" (p. 129). But it really is "watch and be watched." The ancient question, *Quis Custodiet Ipsos Custodes?* ("who watches the watchers?") is being answered by "the watched should watch the watchers." This is why the concept of surveillance is not enough: The network breeds the hacker, hypocrisy spawns punk, the corporation creates the whistleblower, governments mandate rebels, and surveillance leads inevitably to sousveillance.

It is appropriate that Steve Mann coined the term (Mann, Nolan, Wellman 2002). Inventor, activist, author, he has been a live node on the net for over 40 years. He is always recording what he sees and what he hears, an eyeborg, uploading it to the cloud, drawing in signals from the ether. But he's not like the antenna men of Neal Stephenson's *Snow Crash*. They are unthinking extensions of larger organizations; Steve Mann is his own autonomous node.

Different types of "veillance" proliferate: dataveillance; artveillance; contraveillance; self (soi? auto?) veillance; and watching equals—equiveillance. Or, more amusingly, McVeillance, named for one of Mann's confrontations with other network nodes, in this case a McDonalds in Paris, where workers assaulted him for filming and then lied about it, even though he had streamed the attack live (Mann 2012). And when we write about veillance? Métaveillance, obviously. Second-order veillance.

We live in Veillance Society.

Reality is information, and humans hunger for it. That hasn't changed now that we catalog a small bit of it electronically. But digital surveillance information is usually modified on purpose or by processing. This can be benign or deeply misleading, as is the fetal "heartbeat" of the first trimester so beloved to Republicans and other forced birthers, that is often played to excited parents. It turns out that that "heartbeat" is really an artifact of the medical surveillance algorithms used to analyze the pregnancy. One of them turns faint electrical activity in the forming fetal heart into an audible beat. But there isn't an actual heart yet, just some developing cardiac tissue. The heart doesn't come until much later, but that doesn't keep idiots from playing the "beat" at legislative hearings or saying they "heard" the heartbeat when what they heard was the computer making a heartbeat sound. There is no such thing as a fetal heartbeat at six weeks of pregnancy (Harvey 2022).

Information is also about attention, how much attention can we bring to bear, and that takes us to veillance. It is a special type of information flow, from the subject to the observer. Why is cyberpunk a good guide to this? Because it takes actual technoscience seriously but suspiciously, knowing it is about power. A useful life lesson. Also, it is amusing and always offers at least a glimmer of the dream that all is not lost. We have humor and hope, both necessary to survive.

The Panoptic Present

In the late 1970s I worked as handyman/secretary at Bing Nursery on Stanford University's campus. It was more than just a pre-school, it was also a research center for Stanford's Education and Psychology Departments. All the classrooms had hidden observation rooms with one-way mirrors and mic'd ceiling lights so that graduate students and faculty could monitor the children's behavior, including vocalizations, and record it on tapes and in log books. Different research studies looked for different things, but there was almost always someone watching.

When toys broke I was the one called into the classroom to fix them. Being one of the few men around, and adorned as I was with work boots and a tool belt, I was a popular visitor. One day, early into the job, I was told that a big red wooden truck had a damaged wheel so I went to fix it. A small squad of little boys gathered around me as I went to work. At one point my screwdriver slipped, cutting slightly into my hand.

"Fuck!" I said, and one little boy grabbed my arm and pointed at the light fixture above us.

"Be careful," he hissed. "The lights can hear us."

It turned out the toddlers knew about the one-way mirrors and the microphones but kept that knowledge secret from the adults, passing it on, older to younger, as they cycled through. I kept their secret, even though it called into question hundreds of studies predicated on the little subjects not knowing they were under observation.

I lived in a world of protests, our demonstrations were filmed, our radical groups infiltrated, and I spent time in jail. We took care in our letters and knew our long-distance calls were searched for key words. We expected informers, infiltrators, agents provocateurs...but my first intimations of the spread of ongoing, mass technical surveillance came from my experience at Bing Nursery School. It was a panopticon.

Panopticon—A prison designed by Jeremy Bentham for his prison warden brother. Tim Jordan, a sociologist, explains it was...

...a round hollow building with only a tower at its centre. The outside wall was one cell thick, so that the outside window of each cell allowed light in and the inside window faced the inner tower. This meant every occupant of every cell was isolated from each other but became a silhouette to the central tower. (1999, p. 200)

Michel Foucault used the panopticon to explain how maintaining control in society has shifted from using mainly hard power (police, soldiers) to soft (social workers, doctors, accountants). He pointed out that the "supervisor" could peer into any cell and see "a madman, a patient, a condemned man, a worker or a schoolboy." But that wasn't all, because of the design, backlighting would silhouette each cell occupant, making them "constantly visible." He concluded, "Full lighting and the eye of a supervisor capture better than darkness, which ultimately protected. Visibility is a trap" (Foucault 1979, p. 200).

And what better trap than cyberspace, where "mechanisms" most certainly arrange "spatial unities that make it possible to see constantly and to recognize immediately"? Jordan notes that in this system everyone was "watching someone, who is being watched and watching someone else." He agrees with Foucault that this "form of power" spread to "all institutions of modern societies," concluding "The coming of cyberspace seems to many to offer greater possibilities for panoptic mechanisms to make more corners of society visible." Of course, it turns out this isn't always a good thing, the digital world makes surveillance much more effective—"techno-hopes crumble into techno-fears with a simple shift of perspective" most easily in cyberspace (p. 201).

Cyberspace is the cyberpunk's natural environment, as William Gibson (1987b, pp. 16–17) proclaimed:

We're an information economy. They teach you that in school. What they don't tell you is that it's impossible to move, to live, to operate at any level without leaving traces, bits, seemingly meaningless fragments of personal information. Fragments that can be retrieved, amplified...

It was imagined many times before it existed: Memex, The Second Plane, The Grid, The Matrix, OASIS, The Virtual World, Metaspace, Cyberspace, The Web, The Net, The Interweb and now META. While still lacking faux embodiment and neural links, it is here. Shaped by our dreams as much as our technologies, with a Dark Web, corporate intranets, and MILNETs (military networks), it is where billions of human consciousnesses spend trillions of hours. Marshall McLuhan realized in 1966 that we now live in an "instantaneous world of electric information media." He was particularly struck by its "all-at-onceness." It has the allure of synchronicity with its instant gratification. In typical McLuhanesque exuberance he proclaimed: "Time, in a sense, has ceased and space vanished." He concluded that "like primitives" we now live in a "global village" vitalized by this movement of "instant electronic information" and like all villages "everybody is maliciously engaged in poking his nose into everybody else's business" (McLuhan 1966, p. 46).

McLuhan was quite conservative and didn't see the attraction. It turns out this cyber place isn't that much like a village anyway because it is heterogeneous to the limits of imagination and anonymity is still possible, if care is taken. Just as with the postmodern urban environment, there are blind spots, disguises, bolt holes, and temporary autonomous zones that allow the marginalized, especially punks in the widest meanings of the word, to recreate and work. But surveillance is spreading thanks to viral AI; some think it could become perfect. That dream is a "soft" version of the actual, the hard, panopticon.

In talking about the transition from control mainly through hard power to regimes based more on soft power, Foucault says they need:

[P]ermanent, exhaustive, omnipresent surveillance, capable of making all visible, as long as it could itself remain invisible. It had to be like a faceless gaze that transformed the whole social body into a field of perception: thousands of eyes posted everywhere, mobile attentions ever on the alert, a long hierarchized network. (1979, p. 214)

This is what surveillance capitalism (and other forms) has produced. The danger is what Jordan calls, "Total Surveillance or the Coming of the Superpanopticon," raised by "The doomsayers of cyberspace" (1999, p. 197). It is a "nightmare" of total invisibility, and cyberspace...

...is the ultimate system of surveillance, the ultimate tool for repression and the nightmare of totalitarian societies in which not only is everything watched and recorded but any action considered out of the normal is a reason for investigation. The ghosts of all liberal democratic societies return here and in virtual whispers repeat the fear known as 1984 and Big Brother. (pp. 199–200)

Gibson noted, "the street finds its own uses for things" (1987d, p. 189) and those things can in turn end up shaping the streets. In real cyberspace there is resistance, there are bugs, there are gaps. In this, cyberspace mirrors physical reality and vice versa.

Count Zero, aka Bobby, is a young hacker in the eponymous Gibson novel. He is trying to figure out how cyberspace relates to the physical world when a Priest/Crime Lord named Lucas enlightens him by telling the story of the priestess Jackie, who channels an "icebreaking" (decryption) program from the Voodoo God Denbala (actually an AI) while jacked into cyberspace.

"Think of Jackie as a deck, Bobby, a cyberspace deck, a very pretty one with nice ankles," Lucas grinned and Bobby blushed. "Think of Danbala, who some people call the snake, as a program. Say as an icebreaker. Danbala slots into the Jackie deck. Jackie cuts ice. That's all."

"Okay," Bobby said, getting the hang of it, "then what's the matrix? If she's a deck, and Danbala's a program, what's cyberspace?"

"The world," Lucas said. (1987a, p. 114)

In this view, a growing perspective in 21st century postmodernity, meatspace (IRL—In Real Life) is a subset of cyberspace. It is all flows of data.

Veillance and Viruses

The contradiction has become an integration.

—Bruce Sterling (1986, p. xii)

Every technology has its original sin.

—Pat Cadigan (1991, p. 435)

In Veillance Society corporations are ascendant. Nation-states aren't gone, usually, but more and more they are instruments of business alliances, *zaibatsus* in cyberpunk parlance, borrowing from the Japanese. Decades before Citizens United, a Gibson character remarked, "The blood of a zaibatsu is information, not people. The structure is independent of the individual lives that comprise it. Corporation as life form" (1987c, p. 106).

These "life forms" don't just want cheap labor and resources, guaranteed markets, and the outsourcing of costly externalities like climate destruction. They want your loyalty and more. Facebook/Meta, Amazon, Alphabet/Google, and many others run neuromarketing experiments watching your online behavior, all the better to know you for their ends. Foucault (1979, p. 204) pointed out that, "The Panopticon is a privileged place for experiments on men, and for analyzing with complete certainty the transformations that may be obtained from them." In cyberspace this is literally true.

Shoshana Zuboff (2019) sees this as part of the spread of "surveillance capitalism." But she warns us, watching isn't the ultimate goal it is just in first step needed "to change people's actual behavior at scale." It is "a parasitic form of profit" that in the end produces a "profoundly anti-democratic power." Yet it is not just about big data from social media, it includes scientific and technical knowledge, which is also fungible.

The antihero of Gibson's "New Rose Hotel" ponders this while on the run from his employers, the zaibatsu Hosaka. He has failed to successfully kidnap Hiroshi, a scientist from a rival group, and instead led several key scientists from Hosaka into a deadly trap in Marrakesh. Now he is being hunted by his disappointed clients. Reflecting on the value of his target, he realizes that Hiroshi's "edge" is:

Radioactive nucleases, monoclonal antibodies, something to do with the linkage of proteins, nucleotides... hot proteins. High-speed links...Hiroshi was a freak, the kind who shatters paradigms, inverts a whole field of science, brings on the violent revision of an entire body of knowledge. Basic patents, he said, his throat tight with the sheer wealth of it, with the high, thin smell of tax-free millions that clung to those two words. (1987c, p. 108)

New knowledge is almost always a valuable "edge" and this includes all the data sucked up by the ubiquitous surveillance systems of today, only recently perfected.

Veillance Society started to become real with the first U.S. census run with proto-computer machines. Then governments built digital computers to crack codes in war and to build weapons of mass destruction. It grew with the massive collection of social big data, including secrets, by governments, which eventually provoked the widespread revelation of secrets through Wikileaks, Manning, Snowden, and others. It became dominant when corporations built social media platforms and mobile devices that took Facebooking and Snap-chatting and Instagramming to almost every corner of society, producing an infinite flow of digital data.

But no matter what happens with digital information, at least we will always have our private thoughts deep and secret in our minds. There we can think whatever we want and no one can know. Right? No. Or at least, not for long. Just as digital technoscience is growing a Veillance Society right before our eyes, sucking up secrecy, it is also creating the technologies to read, and thus inevitably control, human minds.

Such as optogenetics, a key approach for new mind-healing to mind-reading to mind-writing technologies. Optogenetics, *the* hot new technoscience, involves genetically modifying creatures (so far flies, worms, rats, and nonhuman primates) so that their brain, even on the level of one neuron, can be manipulated by light of different colors. With some genetic modifications blue light activates neurons, with others orange light depresses their activity. Already scientists can "read" a great deal of what a person is thinking thanks to generative AI learning the neurological terrain. Soon, this technology will make direct mind control possible (Gray 2014b). This watching of consciousness we can call *penséeveillance* or, perhaps, *veillance de esprit*.

It is all about what you see in the brain, and what you can make the brain "see." In *Count Zero*, William Gibson describes the experience of the mercenary Turner while his consciousness is being rebuilt by a Dutch neurosurgeon, after a bomb from a disappointed employer destroyed most of his body:

> He spent most of those three months in a ROM-generated simstim construct of an idealized New England boyhood of the previous century. The Dutchman's visits were gray dawn dreams, nightmares that faded as the sky lightened beyond his second-floor bedroom window. You could smell the lilacs late at night. He read Conan Doyle by the light of a sixty-watt bulb behind a parchment shade printed with clipper ships. He masturbated in the smell of clean cotton sheets and thought about cheerleaders. The Dutchman opened a door in his back brain and came strolling in to ask questions, but in the morning his mother called him down to Wheaties, eggs and bacon, coffee with milk and sugar. (1987c, pp. 1–2)

This isn't a government doing this, but a corporation. In *Snow Crash* the Federal Government has been reduced to one of many power systems, somewhat inferior to the CosaNostra Pizza Empire and various other criminal/corporate and religious mind control cults.

Everyone is trying to watch everyone else. At times it is hard to tell who has power over whom. Even when you are the gaze you are part of a relationship, as Gibson's mercenary Turner, discovers:

> The *intimacy* of the thing was hideous. He fought down waves of raw transference, bringing all his will to bear on crushing a feeling that was akin to love, the obsessive tenderness a watcher comes to feel for the subject of prolonged surveillance. (Gibson 1987a, p. 24)

The watcher and the watched, the torturer and the tortured, the killer and the killed, it is always intimate whether both sides of the relationship realize it or not. Veillance is about the flow of information in one direction, as in *surveillance* from above onto those below. After all, many people think God watches over us. In the military, to see combat from above is to use "God's Eye." But Veillance Society has no god, and Sauron's one eye is not enough. It is now a world of eyes but not just Big Brother's. All can watch, and act.

William Gibson clearly sensed how complicated this would become. In 2003 he wrote about the 50th anniversary of George Orwell's dystopia *1984*.

I say 'truths,' however, and not 'truth,' as the other side of information's new ubiquity can look not so much transparent as outright crazy. Regardless of the number and power of the tools used to extract patterns from information, any sense of meaning depends on context, with interpretation coming along in support of one agenda or another. A world of informational transparency will necessarily be one of deliriously multiple viewpoints, shot through with misinformation, disinformation, conspiracy theories and a quotidian degree of madness. We may be able to see what's going on more quickly, but that doesn't mean we'll agree about it any more readily.

1984 is when he published *Neuromancer*.

But the world in most cyberpunk isn't dystopia, yet. It isn't a great world, it is noir, but there is room for agency in the cracks. There is joy and political change. It isn't *1984*. What it is, however, is deeply flawed, with rapacious corporations, corrupt governments, and organized crime linking them all, and always overhead looming existential threats like climate collapse, nuclear war, and neurotic AIs threatening everything. In many ways this is our world, except the AIs are not yet conscious even if they do hallucinate. We are on the verge of disaster, as the cyberpunk worlds are, but we aren't there yet.

The power of corporations is at the heart of the problem. Most mobilize every advantage they have, from micro-monitoring their employees (keystroke by keystroke) to manipulating consumers at every turn. Thirty years ago Mark Dery (1996) warned of a fundamental and spreading alienation. He used the experience of Visual Mark, a synner ("virtual reality synthesizer") from Pat Cadigan's *Synners*: "Surveillance cameras everywhere in the twenty-first-century L.A. of the novel, become his eyes, computer console speakers his voice boxes, the measureless vastness of global cyberspace his dominion" (p. 253). For Dery it is an "original syn," a move from the living cyborg to the bodiless fantasy of the deeply alienated (p. 253). But actually it is the impossible dream of an escape from our actual cyborg existence. Yes, many live much of their lives in cyberspace, but you need your body. Without your body you are dead. That's not going to change, not even for the machines.

Some people bristle at the ubiquitous pandemic claim that "We are all in this together." "No we aren't," they say. They do have a point. The viral underclass, those who have to work, the poor, the homeless, LGBTQIA2S+, Native Americans, Latinos, Asians, and African-Americans all suffer disproportionately. Still, in the final analysis, we are all in it, even if the 1% don't actually believe it—despite repeating it relentlessly through their corporations and sinecures up on the commanding heights of society. Because while a very few can hide for a time from the direct consequences of pandemics, and even from its underlying causes, in the long run none of us can hide from pandemics, let alone from fascism, total war, or ecosystem collapse.

At the end of *The Plague*, Dr. Rieux loses one last friend, his newest, the mysterious Tarrou, who dies in the last days of the epidemic "without their friendship having had time to enter fully into the life of either." Rieux reflects on how Tarrou would have said he "lost the match" to the plague. But the doctor wonders, what has he won by surviving?

No more than the experience of having known plague and remembering it, of having known friendship and remembering it, of knowing affection and being destined one day to remember it. So all a man could win in the conflict between plague and life was knowledge and memories. (p. 291)

But Dr. Rieux thinks about it more and decides that knowledge and memory were valuable.

Nonetheless, he knew that the tale he had to tell could not be one of a final victory. It could be only the record of what had had to be done, and what assuredly would have to be done again in the never ending fight against terror and its relentless onslaughts, despite their personal afflictions, by all who, while unable to be saints but refusing to bow down to pestilences, strive their utmost to be healers. (p. 308)

For Camus, a revolutionary, an existentialist, a philosopher, success in life is measured not by victory over plagues such as cholera or fascism, but whether or not in the struggle you kept your honor.

There are 525,600 minutes in a year. That's around how many Americans died in the first year of the pandemic. A death a minute. On March 12, 2021 I was a week past my 2nd Moderna vaccination. Getting toward immunity, I reckoned. That was 525,600 minutes of lockdown, not totally over yet. Still masks outside and limited contacts, but not isolation. We are never fully isolated, we can always act together. Indeed, the hardest most important projects, such as fundamental political change, can only happen collectively.

I have openly called myself a revolutionary since I was 19. It is not always an easy revelation to make to other people, provoking as it does various disagreements with fellow workers (liberals in particular), or confrontations with police on the line or decisions by hiring committees (although I've won over a few of them) or university administrators (not so lucky there) for nice academic jobs. It isn't just being true to who I am, it also seems to me society needs to know some people believe in the necessity and possibility of fundamental social change. What is hard to communicate to others is how it feels to commit to revolution—It feels like possibility, it feels like hope, it feels like justice. It doesn't mean I am right.

Fascists are revolutionaries too, after all. I had a friend write me recently that fascism seemed more likely in our future than revolution. He's a thoughtful progressive guy but somehow the reality of revolutions—which are as inevitable as thunder storms, as Emma Goldman liked to say—and that they could go either way so-to-speak, eluded him. So what separates fascism from anarchism and feminism? Hierarchy is part of the answer, resisting it or in the case of fascism a desperate need for it. Patriarchy, of course. But there are also deep psychological differences. Feminism which fits well (perfectly for me) with anarchism is impossible in fascism, which is proudly misogynist.

The wonderful activist and author Barbara Ehrenreich captured much of how I see things in her introduction to *Male Fantasies*, the insightful analysis of the psychosexual politics of fascism in Germany by Klau Theweliet. She found hope in the twisted dread of fascism.

It is Theweleit's brilliance that he lets us, now and then, glimpse this other fantasy, which is the inversion of the fascists' dread: Here, the dams break. Curiosity swims upstream and turns around, surprising itself. Desire streams forth through the channels of imagination. Barriers—between women and men, the 'high' and the 'low'—crumble in the face of this new energy. This is what the fascist held himself in horror of, and what he saw in communism, in female sexuality—a joyous commingling, as disorderly as life. In this fantasy, the body expands, in its senses, its imaginative reach—to fill the earth. And we are at last able to rejoice in the softness and the permeability of the world around us, rather than holding ourselves back in lonely dread. This is the fantasy that makes us, both men and women, human—and makes us, sometimes, revolutionaries in the cause of life. (1989, p. xvii)

That's the key to look for—life vs. death. Death is necessary, of course, but it will come whatever we do. We need not worship it. But to survive we do have to valorize life, and find new ways to honor it.

The Affordances of Our Cyberpunk World

For me, the best thing about Cyberpunk is that it taught me how to enjoy shopping malls, which used to terrify me. Now I just imagine the whole thing is two miles below the moon's surface, and that half the people's right-brains have been eaten by roboticized steel rats. And suddenly it's interesting again.

—Rudy Rucker (1986)

The most difficult lessons to be learned are those that come when reality—*in the form of a virus, of our vulnerability to it, of our inadequate governing responses to it—crashes through comforting illusions and ideologies.*

—Benjamin Bratton (2021, p. 1)

Every object, every bit of information, affords certain uses and not others. Different affordances in information technologies (in architecture, code, and protocols) match with different technocultures because of their epistemologies. Is knowledge the result of tradition, elites, and formalisms or innovation, collectivity, and dynamic processes? How one answers this question has clear political implications as well. Contemporary horizontalist social movements incorporate new social technologies into their praxis, into their attempts to change society, while hierarchies seek to reify systems of domination.

Human culture has structures. Information, energy, and matter all flow through human society along established lines. But it is messy. Marriage and property and power rules and rituals are sustained only through human acceptance. Information flows through formal computer networks as well, but even though they grow and grow together and change continually they are maintained by more than belief. They are instantiated in machines and code, organized by protocols.

MILNET (the U.S. military's intranet) is formal, hierarchical, centralized. This is true of all but a few intranets (internal networks). A decentralized network does not have one hierarchy, it has multiple trees or other forms linked together in a meta-hierarchy.

A distributed network has no top, no center. It may contain smaller hierarchical nodes, but its overall form is of a mesh. This is the worldwide Interweb. Facilitated by open committees of experts who set the intercommunication standards, the Interweb has evoked a widely shared ethos, articulated in the 1997 "One Planet, One Net" manifesto of Computer Professionals for Social Responsibility. It declared that "there is only one net" that should be available to all and we are its stewards, not owners. No "individuals, organizations, or governments should dominate." It "should reflect human diversity, not homogenize it." People have a right to communicate and to privacy.

These values don't come from hardware or software. They are political choices. In 1973 Colin Ward published *Anarchy in Action*, where he argued that liberation emerged from principles such as "spontaneous order" and "harmony through complexity." The next year Ted Nelson's *Computer Lib/ Dream Machines* came out claiming computers could foster similar values in society. Since then there has been a proliferation of principles, processes, and even products that link anti-authoritarian social movements and cutting edge computing.

Kevin Kelly echoes Ward when explaining how systems out of external control (weather, economies) regulate themselves. They are distributed with autonomous subunits and high connectivity (1994). But distributed systems are not automatically democratic, look at QAnon.

The affordances of digital information have led to a proliferation of watching. The power of veillance information is the power of the gaze. Andrew Ross argued years before Zuboff did, that information is "a new kind of commodity" that is now "the essential site of capital accumulation in the world." The massive collection of data "converted into intelligence" creates surplus value.

This surplus information value is more than is needed for public surveillance; it is often information or intelligence culled from consumer polling or statistical analysis of transactional behavior, that has no immediate use in the process of routine public surveillance. Indeed, it is this surplus bureaucratic capital that is used for the purpose of forecasting social futures, and consequently applied to the task of managing the behavior of mass or aggregate units within those social futures. (Ross 1991, pp. 126–127)

But watching doesn't just produce "surplus bureaucratic capital," it can be instrumentalist power, making a killing in the market or on the street or in someone's soul. It can expose hypocrites and send the powerful to prison. It can reveal the crimes of empire. It can change the discourse rules and meta-rules that govern society, transforming the alluring spectacle of bread and circuses into a hunger for spectacular social change. Some even imagine that it can lead to the abolition of secrecy.

In his 1990 novel, *Earth* (1990), David Brin describes a near future where a worldwide revolutionary movement has abolished privacy. Their argument is that regular people had long lost their privacy to a surveillance regime of government spying, corporate market research, and the Interweb of things. Only the rich had privacy, and they were using it to steal and cheat and basically fuck over everyone else. The revolutionaries, workers, and the poor from around the world, but especially from colonized/looted countries, wage a successful war on Switzerland to get back their stolen wealth squirreled away in the Alps.

This leads to a society where every moment of every person is accessible to viewing, just as with those who have gone "transparent" in *The Circle*. Of course, most of the time the average person isn't being observed by humans, but they could be. They are always being recorded and watched by computer programs. But the rich and powerful, and the famous and notorious, are under constant sousveillance. The book has other themes—ecological disaster, AIs run amok, the typical SF/CliFi mix—but Brin's take on the oppressive nature of privacy hit a nerve and he ended up defending it in a *Wired* article and a nonfiction book: *The Transparent Society: Will Technology Make Us Choose Between Freedom and Privacy?* (1998)

Brin and Eggers are clearly being provocative, but their insight is real. Digital technology empowered by machine learning is fundamentally changing the way we think about secrecy, privacy, and freedom. The first impacts of this have been mixed. Digital harassment, trolling, and bullying are problems, but the widespread sousveillance of police actions has fueled the growing movement for police accountability. But while Brin's vision is liberatory, Egger's transparent society is anything but. The sousveillance in Brin's fantasy is crowd sourced and decentralized; the megacorporation The Circle seeks nothing less than one seamless transparent corporate society—a cyber Brave New World. It is an attractive idea to many in power.

The COVID-19 pandemic led to much greater surveillance by governments and corporations. Some of it is clearly useful: the widespread monitoring of waste for signs of the virus, Twitter and Facebook eliminating some false and dangerous claims about the pandemic, government investigations of pandemic related frauds. Some of it debatable: enforcing mandates around performative precautions such as massive disinfecting or mandating masks at lonely beaches and policies more in pursuit of the surveillance state than ending the pandemic. In "The Virus of Surveillance: How the COVID-19 pandemic is fueling technologies of control," Félix Tréguer (2021) warns us that the "war on the virus" is strengthening an "authoritarian liberalism" that is using the crisis to protect "industrial capitalism" while eroding society itself. This is even more true of authoritarian illiberalism.

The clearest proof of this is China's Sesame program, where every Chinese citizen is "scored" by their credit, their adherence to the law, their comments on the regime, the ratings of their friends, and a number of other metrics. In the Moslem province of Xinjian every male is having his DNA recorded and in numerous places extensive face recognition on the street is being used (Hvistendahl 2018). This system was strengthened by the COVID-19 pandemic. China's Zero COVID policy wouldn't have had a chance without the thick surveillance of machines and humans on the population that was already in place. The Chinese government added to it with a mandatory COVID app—necessary to enter public buildings, use banks, ride public transport, or leave their city—and then used it to quarantine people in Henan who protested the freezing of accounts at four rural banks that had run out of cash (Wong 2022).

There are dozens of projects in China aiming to integrate all the different data on citizens into a "one person one file" system, making it more effective (Baptista 2022). Yet it is not enough, and even as they tried to implement increased AI to analyze the torrents of data they collected from a billion people and reign in the pandemic, Omicron-2 escaped into Shanghai and broke out in different Chinese cities, leading to protests that shattered the lockdowns and led to their termination by the regime. Despite that retreat, contagious viruses almost always lead to more surveillance. It is disaster authoritarianism.

This isn't just the mania of authoritarian regimes. In the Netherlands, "smart city" initiatives have led to the collection of a wide range of data from people out on the streets, with the goals of reducing night crime, figuring out why people come downtown ("personal mobility profiles"), roaming scanner cars to give tickets and target tax debtors, and so on. While the data is "randomized," it is easily re-individualized (Naafs 2018).

No surprise to cyberpunks; we know all technology is potentially dangerous. Pat Cadigan had one of her heroes proclaim:

All *appropriate technology* hurts somebody. A whole lot of somebodies. Nuclear fission, fusion, the fucking Ford assembly line, the fucking airplane. *Fire*, for Christ's sake. Every technology has its original sin." She laughed. "Makes us original synners. And we still go to live with what we made." (1987, p. 435)

Original sins are baked into the affordances of the technology, but they can also empower hackers and disruptions. This is a key insight of cyberpunk and why it cannot die. It is the yin to The Circle's yang. Tech makes their surveillance more powerful, and it empowers our souveillance.

Bruce Sterling noted, "Anything that can be done to a rat can be done to a human being. And we can do most anything to rats. This is the hard thing to think about, but it's the truth. It won't go away because we cover our eyes" (1999). But when rats look back, their only option is flight. Humans aren't rats. Hackers and activists can gaze back, and challenge the power dynamic.

As the French say: *detournement*. Turn it. Turn it back on them, as the cyberpunk with the dragon tattoo did, as every other fictional cyberpunk hacker does. Turn it, as the Zapatistas turned corporate globalization into a confrontation with international civil society. Turn it, as the Arab Spring, Idle No More, Los Indignados, Occupy, Black Lives Matter, #Woman's March, #Metoo, turned social media on the powerful. But watch yourselves! Of course, most revolutions fail, but as long as the World keeps turning things are still very much up-for-grabs. It ain't dystopia yet.

8

Viral Futures—New Abnormals

No Normal

Today, we are again in a situation where politicians will tell us that what we just went through was a dream, that it's now time to wake up and return to normal. But, actually, 'normality' was a dream; the crisis is reality, for through it we discovered again what is necessary and what is not.

—David Graeber (Graeber and Kantar 2020, p. 224)

The hawk is aerial brother of the wave which he sails over and surveys, his perfect air-inflated wings answering to the elemental unfledged pinions of the sea.

—Henry David Thoreau (2016, p. 120)

On the last day of 2021 I drove to Safeway to do a big shop. I go there five or six times a year, especially if I'm buying spirits. I was shopping for four—Myself, J visiting, and in quantity for my son C and his then girlfriend B, starting a ten-day quarantine. They got COVID again. Maybe in Santa Cruz but probably visiting Huntington Beach for the holidays. It was like a bad flu for fatigue and aching, C reported to me, but nothing in the lungs. They needed vegetables and whisky. I left it at the foot of my stairs and waved to them from my balcony as they picked it up.

At the store I tried to wear an N95 mask I'd bought when the pandemic first hit, but I just could not bear to put it on. Not that I don't dislike my double cloth theme masks (Doctors Without Borders, Golden State Warriors, UCSC Slugs, Stanford) but that N95 mask was way more hateful. COVID-19 taught me a great deal. For one thing I learned how much I now hate wearing masks, despite having worn them often in my blue collar years—woodworking, painting, insulating. When I used to blow cellulite or fit fiberglass bats into attics that were as hot as 110 degrees Fahrenheit, I wore breather masks and full overalls. Now a little mask bothers me. So I avoided going out.

Hiding from Sartre's hell ("other people"), I discovered serious bird watching. My balcony overlooks a veritable forest with at least 15 different trees, mainly eucalyptus and redwoods. Without noticing I became a birder. I've always had a thing for owls, and I was fond of birds in general, fierce little dinos, so much like the feathered microraptors, *dromeosaurids*, of 120 million years ago. The continuity is comforting. I adore the hummingbirds, frequent visitors to my many red flowers. Many times during lockdown, especially when C was working 16-hour days torturing grapes into wine, my only real interaction with another creature was the intense living vibrations, a joyful heavy buzz, of a hummingbird feeding next to my head as I puttered in my balcony garden.

Mainly I watch crows and hawks. They have a complicated relationship. Hawks hunt other birds, of course, but crows hunt as well, especially chicks, even baby hawks. They hate each other with much cacophonous animosity. Usually crows bother hawks until the hawks get annoyed and drive them off. Much as I like the gregarious crows and their solitary big cousins the ravens, it is hawks I love. There is something about their grace and fierce calm that soothes me. Watching them take every moment as the precious gift it is, to be accepted never to return, is a consolation. They give me the courage to reject false hopes, like the fantasy of the normal.

The new normal is there is no normal. It is all abnormal from now on! Stop whining, this is postmodernity. Even the "normal" of modernism was relentless change. With apologies to Heraclitus, we can never step into the same pile of dog shit twice. Each time it is different...Sure, the only constant is change, after all. *But this is different.* Humans have massively messed with Earth, called it culture or even civilization, and it turns out it is a change machine. Yes, life is based on change, circle of life and death, cue the Lion King theme, but biological evolution has a leisurely pace. A little natural selection here, a little natural selection there, and in a million years you've got something.

But now it is clever naked apes tinkering with life like we do leather, wood and stone. Artificial selection is so much quicker. 20,000 years and we've made of noble wolves toy dogs fit only for the purses of the rich. And that is only biological evolution.

Cultural evolution is faster still. It has created participatory evolution, not just choosing our mates but also modifying our genome, as in yummy genetic engineering, now even CRISPR! It has worked, but at a cost. Billions of humans have overrun the planet, warming it and driving the Sixth Great Extinction while exposing our overextended selves to new viruses and other pathogens that live in the flesh of wild and domesticated creatures—our new abnormal.

Which isn't normal at all. It won't get back to normal 2019 ever. And was 2019 normal? In what way? No new transforming technologies? No crazy never-seen-before politics? No records set in climate change, mass extinctions, energy consumption, information processing, human population?

The concept of the "New Normal" was first put forward in 1995 in Oklahoma City to describe what it was like in the aftermath of the bombing of the Federal office building by right-wing terrorists that killed 168 people. It was revived after 9/11 to address how it felt to live in a world where such things happen. But such things have been happening for millennium, just not recently in Oklahoma City or Manhattan. These were not new for most of the world. Terror bombings, high security everywhere, epidemics, medical crisis, mass extinctions, relentless technological change…This is the world we live in. This is the world we have always lived in.

In a funding pitch from the Democracy Collaborative, Arundhati Roy wrote there is opportunity in this. She said, we can use the "terrible despair" of the pandemic to "rethink the doomsday machine we have built for ourselves. Nothing could be worse than a return to normality." So no more longing for the old normal or a new normal. Now is the time to focus on what new abnormal we want, in the context of the viral time(s) we are living through. After all, it could be better than anything we've ever had.

Or it could be worse. We may be in a dark timeline, yet it is far from the darkest possible. Five years before COVID-19 began infecting humanity, RHDV2 emerged among domestic rabbits in Europe. Some 70% of infected rabbits die within a few weeks from internal bleeding and liver failure. It has killed millions of rabbits, and hundreds of thousands of their predators have also died from starvation. There is no treatment; there is no vaccine. Not only is it deadly, but it has high transmissibility and hardiness

(Gammon 2020), and it is spreading to the native hares of the UK and Ireland. Diana Bell, a professor of conservation biology at the University of Anglia, sees RHDV2 as only one of a growing number of new viruses and pathogens coming out of the final destruction of the wild that not only threatens rabbits and hares, but us. "It's like whack-a-mole" as farming spreads viruses to the wild and vice versa. She concludes that this pandemic has shown "how little we know about the diseases that are already present in our wildlife" (Barkham 2021).

Therefore, no surprise that in 2022 the worst bird flu in U.S. history swept the country, killing 52.7 million fowl outright or through culling. Spread by migrating wild birds, it had a particular impact on turkeys and chickens, leading to sky high prices and shortages (Chappell 2022). Into early 2023 eggs were both scarce and dear; Santa Cruz even ran out for a few days. By the middle of 2024 the CDC judged the health risk in the U.S. from its spread still at low, but it has been mutating and infecting many birds and mammals and killed a man in Mexico. Bird flu is much more deadly than COVID-19 to humans, but so far its transmissibility is almost zero. But former CDC director Robert Redfield says a bird flu pandemic is inevitable, "it's not a question of if, it's more of a question of when," and it could have lethality of up to 50% (Dicker 2024). We shall see.

Exponential growth is at the heart of such dangers; exponential growth fuels our hopes. We fear the speed of the virus; we hope for speedy vaccine development and production. COVID-19 is a killer but it isn't as deadly as RHDV2. Yet, from the perspective of pathogens, we are no better than bunnies. A deadly pandemic is almost inevitable, with perhaps only an extraordinary vaccine development campaign to prevent it. A viral bioweapons would be even harder to stop. Even if we learn enough from fighting COVID-19 to avoid the fate of the rabbits, it isn't the only ultimate crisis humanity faces as fallout from our own great success.

There are at least three others—general nuclear or biological war, Global Warming/The Sixth Great Extinction and AI run amok. All three are powered by exponential dynamics driven by the relentless expansion of science and technology. Concretely, life by life, what do these threats really mean? All people die. We have other diseases, mass shootings, natural disasters, multiple deaths happening all the time. But a life is not a statistic; every unnecessary death is a tragedy.

Dr. Anna DeForest, a trainee neurologist, was drafted into her hospital's COVID wards. In *The New England Journal of Medicine*, she tells about the death of one of her patients, someone Dr. DeForest never knew since they were intubated before she took the case. It is the husband she knows. "Your wife is very sick, but stably sick," she reassures him. But his wife soon declines. Fluid from her lungs is first "frothy pink" and then "the frank red of blood." Dr. DeForest tells him that his wife is stable, "for now." She asks herself, "Why do I lie?"

The morning her patient dies she doesn't want to be at work. Yet she goes in to "crawl dejectedly into scrubs" and fantasize about what anti-lockdown protesters, almost always white men with guns, would look like "dying on vents" in the ICU. As her patient dies, feet kicking in rhythm to the compressions from "the Thumper" machine, she looks "for hope and find none." When the code is called she feels "I have never mattered less in my entire life." She is grateful the husband is told by someone else. She asks the reader to "forgive me" that "I am not the one who hears him cry out in grief." She wonders, "What else is there to say? You are dead, like so many others, and the rest of us are left to live in the absence of any certainty" (DeForest 2020). Dr. DeForest will certainly never be normal again.

On September 18, 2022 President Biden announced the pandemic was over. It was not, but he was not alone in wishing it was so that we could get back to "normal." Early in 2022, under the rubric "Urgency of Normal," a team of doctors pushed for ending COVID-19 restrictions as soon as possible. One of them, Dr. Prasad, has set something of a record in being wrong about the pandemic, always minimizing it—not as bad as the flu, expose your kids to diseases there are vaccines for—that kind of thing.

He continued in this tradition, labeling COVID-19 endemic before it was, ignoring that endemic can be very deadly. Tuberculosis is endemic, there are cases in 25% of the world; it killed 1.5 million people in 2020 and debilitated millions more. If you are middle class or above you can ignore it, but not if you are in the viral underclass.

A nursing student, Finn Black, writing against the "Urgency of Normal," pointed out that if they mean get back to racism and greed that might be contra-indicated. After all, "Normal meant spending more per capita on healthcare than any other industrialized country while consistently having poor outcomes. Normal is why we are in a crisis right now." Black concluded that if we want to say that being healthy is purely an individual choice and that "high death rates among those with little political power are acceptable" than that might seem normal, but it isn't. A "better future is possible" he reminded us, if we work toward it (Black 2022).

Weird Lessons

85. As waves come with water and flames with fire, so the universal waves with us.

—Reps and Senzaki (1957, p. 205)
(Vigyan Bhairava, Sochanda *and* Malini Vijaya Tantras, *before 2000 BCE*)

I'd wanted to escape history by running to the hawk. Forget the darkness, forget Göring's hawks, forget death, forget all the things that had been before. But my flight was wrong. Worse than wrong. It was dangerous. I must fight, always, against forgetting.

—Helen Macdonald (2014, p. 265)

One of the gifts I received from the COVID-19 pandemic was reading *The Plague* by Camus. I was surprised I hadn't read it already; he is one of my heroes. Written in 1948, ostensibly about an epidemic of bubonic plague that devastates the Algerian French Empire city of Orlan, it is also about how people responded in France to the Nazi Occupation. Camus, for his part, joined the Resistance and edited *Combat!* underground, but most French did not resist, and many collaborated.

Collaborate or not is still the question. Why not collaborate? Cooperate? These are the virtues of civilization. Go along to get along. Play nice. Don't rock the boat. Go with the flow. But the flow is often evil; the flow is often a deluge of dying. The literature of plagues make it clear that the questions raised by mass deaths, by all four of the Horsemen of the Apocalypse not just the pale rider, are those of the human condition. Why do we live and why do we die and why do we know it? Is love as real as death? How should we face the great challenges of our lives? We have to start by not fooling ourselves, or letting ourselves be fooled, by vivid lies and seductive dreams of normality.

David Graeber (2021), soon before his death, warned us to wake up.

Because, in reality, the crisis we just experienced *was* waking from a dream, a confrontation with the actual reality of human life, which is that we are a collection of fragile beings taking care of one another, and that those who do the lion's share of this care work that keeps us alive are overtaxed, underpaid, and daily humiliated, and that a very large proportion of the population don't do anything at all but spin fantasies, extract rents, and generally get in the way of those who are making, fixing, moving, and transporting things, or tending to the needs of their living beings. It is imperative that we not slip back into a reality where all this makes some sort of inexplicable sense, the way senseless things so often do in dreams.

All we have to do is wake up.

Awake we have the power to avoid many future disasters. We can make viruses, change viruses, transform the context and criteria that allow viruses to thrive or not. Viruses propagate and die within systems: biological, digital, social. It is a problem of cybernetics, of control. The many can end the authoritarian control of the few if we control ourselves, while at the same time no longer pretending that we are in absolute control of the vast natural "out-of-control" systems we are part of. To start, we have to control the velocity of the changes we provoke.

Velocity is the dynamic of viral time. This is one of the weird lessons we have learned from COVID. Remember how quickly the pandemic came on; how suddenly so much changed? We need to own this, and use it, because climate change and the dangers of AI and political instability are accelerating; tipping points and singularities (from which you can't come back) are not just coming, they are all around us. The world will never be the same, we have lost parts of it irrevocably.

Acceptance is another pandemic lesson. We had to accept the pandemic or collapse in impotent fear. We have had to accept the reality of Long COVID, too many people have it. I've a half dozen good friends dealing with the fallout of "post-acute sequelae of COVID-19" as it is called on medical reimbursement forms. There are now over 200 different documented symptoms and at least 65 million with the syndrome (Davis et al. 2023).

We have to accept that we don't know everything. Yes, we might discover why diagnosed mental illness increases one's chance of dying from COVID-19. Inflammation seems a crucial clue (Resnick 2021b). Yes, we know, as Canadian psychiatrist Steven Taylor, an expert on the psychology of pandemics, reported that Covid Stress Disorder impacts 15% of people, but will we ever know how to treat it, or the more generalized Coronaphobia, any better than the other psychological afflictions of postmodernity? On the other hand, acceptance is no doubt driving the mental improvements by the pandemic. Taylor also reports "post-traumatic growth" went up in 77% of people, including more altruism, increased resilience to stress, more connection to family and loved ones, and "recognition of new possibilities" (Taylor 2022).

We do need to accept that SARS-CoV-2 will linger long in the body politic. But it will change and we will change. Viruses, human brains, cultures—all are plastic.

Plasticity sets the limits of the possible. COVID-19 revealed that we live as if some things that are actually transitory are forever. Going to work, going out dancing, capitalism, are all options that can be undone by events, not inevitabilities we can count on. It turns out not only can most people imagine the end of capitalism, we can imagine the end of work as we know it, life as we know it. We can, and must, live lives we couldn't imagine before. We must imagine in new ways, beyond old clichés.

Bill Gates said to the Massachusetts Medical Society in 2018 that, "The world needs to prepare for pandemics in the same serious way it prepares for war." But he was wrong. He wasn't just talking about the scale of the effort (which, considering the amount spent for war would be too much, actually). He meant in the confrontational, tactical, hierarchical, science-obsessed, elitist, reductionistic ways we prepare for war. We prepare for war as a profit-making enterprise. It is not just the military–industrial complex President Eisenhower realized was the greatest threat to Western democracies, but also the kleptocracy of Russia and the militarized economies of Egypt and China, where soldiers run many of the most lucrative industries, and indeed most of the world where economies and nation-states are organized first and foremost for war, and profits from war. Profit is a problem, not a solution.

We won't survive climate change if we act like we are at war with the Earth itself. We need to be in solidarity with the Earth. When we struggle for a just and sustainable world, we aren't doing it "for" nature, *we are nature fighting back.*

Solidarity is not about pity or liberal guilt. It is about recognizing who we really are. Individuals—despite being wonderful and unique—are nothing by ourselves. Alone we cannot exist. There is society and it gets sick. We know in most disasters people pull together. I've seen it in Santa Cruz during our fires, our floods, our earthquakes, and with the pandemic and the rise of American fascism. But Sonia Shah (2016, p. 118) thinks pandemics don't provoke the social solidarity of other disasters.

Unlike acts of war or catastrophic storms, pandemic-causing pathogens don't build trust and facilitate cooperative defenses. On the contrary, due to the peculiar psyche experience of new pathogens, they've more likely to breed suspicion and mistrust among people, destroying social bonds as surely as they destroy bodies.

Steven W. Thrasher agrees with her. He thinks that SARS-CoV-2, "like all pathogens," isn't "we are in this together," a "great equalizer," rather it is "a magnifier of the divisions already present in our world" (2022, p. 5). He is not completely wrong, of course, especially in terms of AIDS, which was politicized in horrible ways. Nor is he completely right. Death makes us equal in the long run, that is an integral organizing aspect of pandemics. This is why Shah is wrong as well. She chronicles many instances of fear driven racism and hatred in the historical response to pandemics, but it isn't hard to notice that there are as many instances of loyalty and love, even for strangers, even at the highest costs.

Anyway, what choice do we have? We have to "Stay with the trouble" as Donna Haraway has long urged. That means, especially, not giving into our tendency, driven by cognitive dissonance, to simplify what can't really be reduced. We have to learn to hold opposing dynamics in tension, and not collapse one into another (Aronson and Travis 2020). This is what speaking Virus is about. It isn't about unlearning how to speak something else, but rather adding a new conceptual world to our existing fluencies.

How will you know if you are starting to see Virus as a language? Well, it is like English. When you see "English" you have to sort it out. Is it what you "put" on a ball to make it harder to hit? Is it the people living south of Scotland and East of Wales? It is a type of horrible breakfast with beans and fish in unholy alliance? Is it the language spoken around the world by scientists and diplomats, and (sorta) by Australians and 'mericans, and…? Is Virus a contagious algorithm? Is it a biological near-living replication system? Is it a wave of hope or hate (or both) sweeping through a nation? Is it a pattern, a way of thinking, a vocabulary that can help us understand, and therefore control, all of these? Yes. Yes it is.

What will our new abnormal be like? We can't know. But more disasters are coming. Donna Haraway (2021, p. 299) warns,

I think that loss is real and accelerating, and there will be no *status quo ante*. There will be no going back to a prior state. The new equilibrium points will be different, and they will be worse in all kinds of describable ways. I'm talking biologically right now. So I think that extinction is real and accelerating, and anyone who thinks that there's a techno fix is in a state of abstract denialism.

As our world accelerates it isn't enough to predict what might happen, we need to shape it.

Possibilities

Nothing is gained without loss.

—Paul Virilio (1999, p. 54)

No fate but what we make.

—John Connor and his mom, Terminator *franchise*

From early January 2020 I charted the spread of COVID-19 and made various predictions about how many people it would kill. I found this calming. It probably gave me the illusion of control, the drive behind much knowledge production. As an academic I suffer greatly from the realization that knowledge is power. So I study power in the context of new technologies and ever-changing science. But I pay close attention to my delusions (for they are legion), for cognitive blind spots that might warp my analysis. A good check is to make predictions and see how they turn out. They might even be useful, if good enough or bad enough. My predictions about COVID-19 were revealing.

Besides the vaccine prediction noted earlier, I assumed from the beginning that shutting beaches was stupid and wiping groceries silly. Shutting borders and restricting travel in other ways only made sense early on, as in China and New Zealand, for most of the world such policies were too little too late. I made pretty good estimates of eventual death totals, about the unequal burden of the pandemic on the viral underclass and underclass nations, and many other related issues. On the other hand, I could not believe Trump could win reelection. I underestimated QAnon, I underestimated the political viral (and pathological). I also have been surprised by the rapid increases in the powers of generative algorithmic intelligence.

In the *The Plague*'s final paragraph there is a prediction. Rieux listens to the "cries of joy rising from the town" at its reopening. No doubt, much like the cries of joy at the liberation of Paris from the Nazis that Camus had heard, for fascism is clearly as much an infectious pestilence as bubonic plague, and also hard to eradicate.

> Rieux remembered that such joy is always imperiled. He knew what those jubilant crowds did not know but could have learned from books: that the plague bacillus never dies or disappears for good; that it can lie dormant for years and years in furniture and linen-chests; that it bides its time in bedrooms, cellars, trunks, and bookshelves; and that perhaps the day would come when, for the bane and the enlightening of men, it would rouse up its rats again and send them forth to die in a happy city. (p. 308)

This is a pretty safe bet, for both biological and political plagues. More specific predictions are much harder but worth the effort. Every prediction generates possibilities.

From the beginning we heard predictions about COVID-19, and most were wrong. That is how we learn. But at times we learn more about the predictor than the problem. Over-optimistic projections on herd immunity have proven particularly revealing. Marty Markary, a doctor and professor at Johns Hopkins, predicted early in 2021 that, "We'll Have Herd Immunity by April." He clearly did not take into account that new variants were likely and that they might escape the immune response caused by getting sick with earlier versions. Later, he defended being wrong about herd immunity, never achieved, and complained other "scientists shouldn't try to manipulate the public by hiding the truth." But his predictions aren't "the truth." His colleagues who worried his predictions weren't accurate enough to share, considering the false hope it would give people, turned out to be right. Confusing one's best judgement with "the truth" is one reason people don't trust anyone who thinks they know more than they probably do (Markary 2021).

Or consider this from earlier, June 2020, when estimates of needed infection/vaccination rates for herd immunity were as low as "smaller than 43%" (Prof. Tom Britton, University of Stockholm) or even 20%. These are not bad people they just did some bad science, basing their conclusion on guesstimates about heterogeneity of susceptibility (the variation between how likely individuals are to get infected) while ignoring the history of respiratory viruses. Mathematician Joel Miller pointed out that even if the herd immunity level needed was 60%, "it might be another 20% that gets infected while the disease is starting to die out" (Hartnett 2020).

In September of 2021 the COVID-19 Scenario Modeling Hub used nine different mathematical models to predict Delta would probably be the last wave. But as it turned out...Omicron. They couldn't imagine Omicron, actually, and started speculating about what endemic COVID-19 might be like (Karlis 2021b). Another reminder, optimists get more done but pessimists are right more.

Early in 2021 Sarah Zhang accurately predicted that herd immunity might never be achieved, "COVID-19 will likely continue to circulate, to evolve, and to reinfect." She even foresaw that transmission resistance would wane way before protection from serious infection. Vaccines aren't to end pandemics but to prevent them from killing hundreds of millions. Millions dead we know we can live with from the flu, AIDS, monkeypox, polio, STDs, malaria, tuberculosis, and other epidemics, including gun violence and car accidents. The flu turns out to be the best model, unsurprisingly as it is closest biologically (Zhang 2021).

Scott Gottlieb argues that as COVID-19 becomes endemic we should develop comprehensive policies that deal with both that and the yearly flu we've learned to accept, even though it often kills hundreds of thousands in the U.S. alone and costs tens of billions of dollars. But are people ready for masking inside every winter from now on? How much will companies and other institutions pay to make their air flow and filtration systems minimize pathogen spread instead of exacerbating it? Better personal hygiene (and NOT going to work sick) and home diagnostic kits with home treatments could be implemented against both endemic threats and worse pandemics to come. It is a rethinking of how to protect the health of the body politic (Gottlieb 2021).

Epidemiologists have long planned for Disease X, a pandemic worse than COVID-19. It was defined in one 2006 survey of doctors and scientists as: seriously sickens a billion, kills 165 million, and costs 3 trillion. They were 100% sure that one would happen by 2050 (Shah 2016, p. 8). Others, such as myself, imagine even worse. Consider that Ebola has killed one-third of the world's gorillas and almost that many chimps

(Shah 2016, p. 27). An engineered virus that targeted humans could be even more deadly.

In the digital-cultural realm we seem to be facing something equally disruptive. Machine learning based on massive reiterative viral pattern recognition exploded into our mass cultural consciousness in early January 2023. Students started using AI for their papers, websites for their news reports, artists started using AI and winning some prizes, and gatekeepers started banning AI art and texts. This is only the beginning.

Kevin Kelly (2023) explains that the pattern recognition process now mastered by AI is "at the core of almost everything we do" cognitively. There is much more to human thinking, of course, but considering that synthesizing it "has taken us further than we first thought" it "will probably continue to advance further than we now think." True enough, considering the power of the social media algorithms spreading QAnon, mobilizing millions to elect Trump twice. The social disruptions we are seeing now are just the beginning. The COVID-19 pandemic is an example of the disruption one biological contagion is capable of, we'll soon see how much the spread of viral AI can change our world.

Living in Pompeii

We had scarcely sat down when night came upon us, not such as we have when the sky is cloudy, or when there is no moon, but that of a room when it is shut up, and all the lights put out. You might hear the shrieks of women, the screams of children, and the shouts of men; some calling for their children, others for their parents, others for their husbands, and seeking to recognize each other by the voices that replied; one lamenting his own fate, another that of his family; some wishing to die, from the very fear of dying; some lifting their hands to the gods; but the greater part convinced that there were now no gods at all, and that the final endless night of which we have heard had come upon the world.

—Pliny the Younger on the eruption of Vesuvius, 79 CE (2008)

Only in silence the word,
Only in dark the light,
Only in dying life:
Bright the hawk's flight
On the empty sky.

—Ursula K. Le Guin, "The Creation of Éa" (1968, p. 51)

I first visited Pompeii in 1972 when the Turkish freighter I was riding from Athens to Barcelona stopped in Naples for half a day. The ghost city was cool and clean from the first Autumn rains, practically empty of visitors and guards, one of the few tourist attractions that was all that was promised. It was haunting. In early June of 2022 I visited again with J. It was our first real trip since COVID-19. Tests required, masks on trains, many venues still closed, but Italy was wonderful all the same. Sadly, it was to be our last trip together. All things end, after all, including love.

Pompeii seemed different at first. Unlike 50 years earlier there were mobs of other visitors, curated displays, a fantastic museum, solicitous staff, water, and food to buy. But the bones of the city were the same, just more on view. Walking the dead streets you can see how much the Pompeians lived as we do. There was street food and brothels, government officials and different religions, commerce, and art. The story is the same. The lesson is the same. We all live under doom. We all live in Pompeii.

When Vesuvius erupted, Pompeii was still recovering from the devastating earthquake of 62 CE. The philosopher (and bureaucrat) Lucius Seneca (the Younger), wrote about it at some length in his *Natural Questions* (1833) which he dedicated to his friend Lucilius Junior, procurator of Sicily during Nero's reign and a native of the city. He wanted to put forward the Stoic argument that fear can be conquered through knowledge, especially knowing that you will die but don't know when. This was accomplished, for Seneca, by understanding that earthquakes had natural causes that we could understand. He writes about the destruction throughout Campania at length, relishing such details as 600 sheep in a field near Pompeii just dropping dead.

For him earthquakes illuminated the human condition,

> Where, indeed, can our fears have limit
> if the one thing immovably fixed, which upholds all
> other things in dependence on it, begins to rock,
> and the earth lose its chief characteristic, stability?
> What refuge can our weak bodies find ? whither
> shall anxious ones flee when fear springs from the
> ground and is drawn up from earth s foundations ? (1833, p. 222)

He goes on in this vein at length before getting into theories on what causes earthquakes, within a reality explainable by the interactions of the four basic elements of earth, fire, air and water, the accepted science of the day. He compares several theories and settles on Earth farts. He has not forgotten about the dead sheep. Not bad considering the limited framework he started from, but he makes several serious errors anyway that observations—his main mode of inquiry—should have precluded. He doesn't see the connections between such phenomena as earthquakes and volcanos, even though the Greek geographer Strabo, writing almost a century earlier, not only connected earthquakes to volcanos but also to tsunamis (1903). Seneca also assumed the danger of earthquakes is everywhere the same. This served his Stoic worldview, for if death is everywhere equally possible, why fight the inevitable? A typical Stoic death trip. He says don't "listen to the people who have bid adieu" to the region vowing "never" to "set foot" there again. No one can promise them "solid foundations in whatever soil they choose" because everywhere in "the world is subject to the same fate."

Which isn't true of course. The people who fled in 62 CE avoided the eruption of Vesuvius 17 years later. It is easy enough to mock his theories, but he thought much more clearly than most. He also bravely denounced gladiatorial spectacles. On a wall in Pompeii near their arena someone wrote, "the philosopher Annaeus Seneca is the only Roman writer to condemn the bloody games," in reference to his argument that the spectacle was immoral and degrading to participants and spectators alike.

Seneca died before Vesuvius erupted so we don't know what he would have made of it. He was ordered to kill himself by his former pupil, Nero, or see his family destroyed. His third death sentence, actually, but the first he couldn't avoid. He met his death nobly, showing that he could take his own advice on dying, offered at the end of his essay on earthquakes.

> Let us fix this in our minds, and constantly remind ourselves, I must die. When? What matter is that to you? Death is a law of nature; death is a tribute and a duty imposed on mortals; it is the remedy of all ills. Whoever now fears it will one day long for it. Giving up all else, Lucilius, make this your one meditation, not to dread the name death. By long reflection make death an intimate friend, that, if so required, you may be able even to go forth to welcome it. (1833, p. 268)

Pliny the Younger, another Roman philosopher–bureaucrat, wrote the only first-person account of the death of Pompeii we have. He watched the eruption from across the bay, and he almost went with his uncle, Pliny the Elder, the famous naturalist and admiral of the Roman Fleet, when he went off in a war galley to save survivors. Witnesses told his nephew that when the Elder was urged to turn back, he said "Fortune favors the brave" and ordered them in. It happens to be a favorite slogan of Mark Zuckerberg (Losse 2018) as well. Even I've got "Fortune favors the brave," in Latin, tattooed on my left shoulder—it is the motto of the O'Flaherty clan of my ancestors. But it isn't always true, after all Pliny the Elder died on a beach in the Bay of Naples, being brave, although his nephew thought his sickly uncle probably did welcome his heroic end.

We know more than Seneca or Pliny knew about earthquakes, volcanos, viruses, and global warming. We knew a pandemic was inevitable, but our governments did little. We know we are killing our biosphere, and our economy is organized to give more and more wealth to the rich who most profit from its destruction. Would Seneca advocate accepting our fate without resistance when we are its author?

Of course, the "we" is not everyone. Some scientists, political alarmists (so extremists such as myself), fringe academics (me again), renegade bureaucrats, and engineers (my father), have been issuing warnings for decades. Just as in ancient times very few people raised their voices against slavery or even gladiatorial spectacles. It was just the way things were.

In 73 BCE, almost 150 years before Vesuvius destroyed Pompeii, a group of rebelling slaves from the Gladiator Training Academy at Capua, set up camp in the cauldron. They were led by the Thracian Spartacus. When the Roman Praetor Gaius Claudius Glaber besieged them, Spartacus and his army climbed down the steep opposite side using vines that grew in the dormant volcano. They came on the Roman army from the rear and shattered it. They went on to defeat a number of other armies before they were destroyed trying to escape Italy a few years later. Six thousand of them, including Spartacus it is assumed, were crucified along the Appian Way between Capua and Rome.

Pompeii is a lesson; it is also a possibility. Human is human, but each age has its voice—its particular new abnormals. In our age it is clear that the system is producing the viral polycrisis that bedevils us. Any solutions that are based on denying that, even from the nicest most expert liberal academic think tanks and billionaire philanthropists, are worse that useless, they are part of the problem.

For example, six months into the Pandemic the Edmond J. Safra Center for Ethics at Harvard University released their "Roadmap to Pandemic Resilience." It called for testing, tracing, and focused quarantines along with throwing money at the economic problems caused by lockdowns. They had detailed plans for labeling essential workers, isolating adults over 60 from the rest of society, and restarting a reorganized economy and other fantasies such as: "Establish a culture of universal mask wearing" (Allen et al. 2020).

A team of scientists and other academics, funded by NGOs, released their "Roadmap for Living with COVID" two years later and it was much the same. It stressed infrastructure for "testing, surveillance and data," investments in vaccines and therapies, better sharing of information, and interventions in the economy and education designed to maintain the status quo, portrayed as the "next normal" (Albarracín et al. 2022).

In his book *How to Prevent the Next Pandemic*, Bill Gates is equally, maybe willfully, clueless. He advocates doing more of what we've been doing but "better." He doesn't worry about misinformation because "the truth will outlive the lies" and the "practice, practice, practice" of trainings and simulations will somehow solve the problem that even when bureaucracies know what they need to do, they often don't do it, as in stockpiling PPE. Gates is a big believer in expertise, but it was experts who didn't think Ebola was a threat to West Africa and it was experts who thought Sars-CoV-2 couldn't be spread asymptomatically because Sars could not. The problems aren't always fiscal or technical, they can be epistemological (Honigsbaum 2022). The experts did not predict Russia attacking Ukraine or a clear Trump victory in 2024. I'm an expert; I am often wrong. But mix what I know with other experts, my friends, and the general public's perspective—democratize expertise (which is what science is, with some weird rituals and heavy on the meritocratic)—and understanding grows exponentially.

The *Vox* reporter Whizy Kim also read Bill Gates' book and she noticed that Gates felt philanthropy alone was the cure for inequality, helped by businesses driven by profit. For Gates, the solution to the inequalities of today is to double down on the system that produced them. As Kim says, "Gates views inequality as an unfortunate misallocation of resources, an oversight where some people just don't get enough of the pie." Of course, for the fourth richest person in the world, who got significantly richer during the pandemic, it is hard for him to see that the system that made him rich (a Harvard kid with a mother on the board of IBM) "could be an engine of misery." But it obviously is (Kim 2022).

Kim recognizes that there are "quasi-religious characteristics of capitalism." It is a sacred hunger. She shows that while Gates counts on Big Pharma to do the right thing, for a profit, the medical sociologist Howard Waitzkin notes that U.S. life expectancy declined between 2014 and 2017 and again from 2020 to 2022. Rich countries hoarding vaccines, connected entrepreneurs getting rich off massive pandemic spending, and so on, is business as usual. It is why there is a viral underclass.

> Covid-19 shows us that no amount of tech or science innovation will prevent crises like Covid-19 unless we address the root of inequality: an economic structure that's tilted so far in favor of economic growth and the already-wealthy that it systematically devalues people on the lowest rungs of the class system while demanding that they bear the highest costs. (Kim 2022)

In the case of pandemics this isn't just having medical responses, protective technology, tests, treatments, and vaccines. We need better institutions. Traditional institutions are too corrupt. Even liberal leadership fails at an astonishing rate. We need more democracy and less of the evil and fanciful systems in place now. We need to reject profit as a social value. We need to reject the human-nature distinction. We need to speak Virus.

This starts with listening. Since most viruses are benign or even good for people, it doesn't make sense to go on a crusade against the viral. But if a virus is unhelpful, destructive, deadly? One limits access to new hosts, applies anti-virals, blocks catalysts and implements other controls. More concretely, what fundamental changes need to happen to protect us from malicious viruses, biological, digital, and cultural?

1) Power to shared expertise; democratize science.
2) No more profiting from evil.
3) Think viral time—from rapid reactions to long-term institution building. It is not about quarterly or yearly corporate reports, or two-year and four-year election cycles.
4) Leverage suffering.
5) An economy based on actual human needs and desires and not mindless consumption and incredible wealth disparities.
6) Which means more democracy.

Why spread democracy? Because each crisis should serve as a vaccination for future crisis. We need to change our body politic so it can overcome existential threats. A good discussion of "pandemic as opportunity" for the health system in the U.S. argued that the pandemic could help improve it in five specific ways: (1) "Rehire public health workers and rebuild resources"; (2) "Get data and labs out of their silos"; (3) "Put primary care back at the center of health care"; (4) "Tackle the racial and social disparities that threaten the health of all communities"; and (5) "Prepare now for a coming mental health tsunami" (Kenen 2021). It all seems reasonable enough, but these reforms run afoul of the for-profit medical system and, even more, call for a sweeping reordering of American culture.

Reforming the health system will never happen without fundamental political change. The criteria for making knowledge cannot be maximizing profits, it has to be social. There is often very little that is rational about how medical policies are made; they often swing wildly between neglect, and panic, and neglect. Caroline Chen shows this in reporting on why the CDC has failed to eradicate congenital syphilis in babies. While very treatable and controlled in many countries (Belarus, Cuba, Malaysia), in the U.S. cases are rising. The politics of funding have been shaped by the original neglect of AIDs (the "Gay" disease) so it now gets six times more than ALL other Sexually Transmitted Diseases combined. AIDS is a major killer still and deserves funding but not exclusively. The CDC is so underfunded and beset by politics it can't even get penicillin to pregnant women with syphilis (Chen 2021b).

Have you ever wondered who actually made world COVID policy? A team of journalists from *Politico* and the German newspaper *Welt* showed that worldwide policy on the COVID-19 pandemic is basically controlled by four large non-government philanthropic organizations which intertwine representatives from other NGOs, governments, WHO, and Big Pharma. The two biggest players are Bill Gates and his extensions and the Wellcome Trust, started by Henry Wellcome, one of the first pharmaceutical millionaires who invented free samples for doctors.

Unsurprisingly, Gates and Wellcome agree that saving the world requires protecting intellectual and material property. The report's critique on this hybrid government–NGO leadership is that it fails to speak Virus well. For example, it "oversimplifies and misrepresents complex concepts like tipping points and tipping point cascades, and regularly equates nonlinear trends with unstoppable, compounding change" (Banco, Furlong and Pfahler 2022). Their commitment to the political status quo no doubt contributes to an inability to think beyond simple extrapolations of how things already are.

It is not a question of science; it's a question of values. Erica Charters stresses that epidemics don't end in just one way—they have to end medically, politically, and socially (Stobbe 2022). As Omicron1 faded in the spring of 2022, more and more people were declaring the epidemic over in the U.S., even though more than 1,000 people were dying daily and son-of-Omicron was already spiking infections and hospitalizations in Europe. They just hoped it was over. But it wasn't.

At the end of January 2023, over 500 people in the U.S. were still dying from COVID-19 every day, four times the toll of an average flu season. China, which suddenly ended its Zero Covid policies in the face of growing demonstrations, soon had over a billion Omicron infections and perhaps as many as 1.23 million deaths, on top of its first population drop ever recorded, of almost a million people in 2022 (Davidson 2023).

On September 12, 2022, I got my fifth vaccination, the Omicron shot. I also got a flu shot, something I've only started doing since the COVID-19 pandemic started. Learning about Long Flu certainly influenced me. Meanwhile, we learn more about SARS-CoV-2, how new variants can reinfect despite vaccinations and previous infections, how the virus is found throughout the body in some autopsies, how it infects almost every bodily system, and how Long COVID works and what it does.

But asking hard questions and getting unpopular answers takes courage. In a *Nature* survey, 15% of the doctors who spoke out on COVID-19 in favor of masks and/or vaccinations or against fake cures or on the origins of the virus, received death threats, and 22% were threatened with sexual and other violence (Campbell 2021). People don't like to get death threats—I don't. I've received dozens, but have always seen them as a sign I was doing the right thing, hard as it is to face that kind of energy. But I am openly a revolutionary; most doctors aren't prepared for it.

You also need hope to keep going. I suffer from a weird curse; I know the suffering of other people is real. All other people. I don't get compassion fatigue; I can't find peace in denial. Like Cassandra I know what most don't want to hear. It can be debilitating; it is exhausting. Dear reader, I want to infect you. Accept I am real. Accept all the human suffering and joys you hear about on the news and from your neighbors is real. During the writing of this book, dear friends of mine have died, love affairs have ended, hidden loves have been revealed, babies have been born to those I love and millions of other humans, strangers, have died unnecessarily. Tens of thousands of Palestinian women and children have been killed in mass technological slaughter guided by viral AI. From targeting customers with ads to targeting the families of journalists or Hamas members in their living rooms was just a very short step (Gray 2025). It is all real. And I know you are real. I feel you are real. I write to you.

Some days I don't see the hawks but I always hear them, bickering with the crows, signaling each other in their long slow gyres riding the thermals as they hunt. They face each day with clarity, not just because of their amazingly powerful amber eyes, but by their nature. Feathery little dinosaurs in harmony with their bodies, their environment, their needs. Without illusions. The great human power of imagination can also be a weakness if we give into denial, if we don't accept our nature, where we are, what we need to do.

References

Abouzeid, Gihan (2021) *A Regional View of the Global Pandemic*, Arab NGO Network for Development.

Aizenman, Nurith (2021) "The mystery of where omicron came from—and why it matters," *NPR*, December 1.

Albarracín, Doloras et al. (2022) *Getting to and Sustaining the Next Normal: A Roadmap for Living with COVID*, Roadmap.

Allen, Danielle et al. (2020) "Roadmap to Pandemic Resilience" Safra Center for Ethics at Harvard, April 20.

Anglen, Rober, Richard Ruelas and Lorraine Longhi (2020) "Fake social media posts incite fear of suburban marauders, rape and murder across the U.S.," *Arizona Republic*, June 4.

Antonelli, Michela et al. (2022) "Risk of long COVID associated with delta versus omicron variants of SARS-CoV-2," *Lancet*, 339/10343, June 18, pp. 2263–2264.

Aronson, Elliot and Carol Tavris (2020) "The role of cognitive dissonance in the pandemic," *The Atlantic*, July 12.

Arthur, W. Brian (2009) *The Nature of Technology: What It Is and How It Evolves*, Free Press.

Aschwanden, Christie (2021) "Five reasons why COVID herd immunity is probably impossible," *Nature*, 591, March 18, pp. 520–522.

Azarian, Bobby (2021) "Neuroscientist explains how fanatical Trump followers could lead us to societal collapse," *Salon*, August 6.

Azhar, Azeem (2021) "The exponential age will transform economics forever," *Wired*, June 9.

Bacigalupi, Paolo (2012) "How cyberpunk saved Sci-Fi," *Wired*, June 20.

Banco, Erin, Ashleigh Furlong, and Lennart Pfahler (2022) "How Bill Gates and partners used their clout to control the global Covid response—with little oversight," *Politico*, September 14.

Baptista, Eduardo (2022) "China uses AI software to improve its surveillance capabilities," *Reuters*, April 8.

Barkham, Patrick (2021) "New 'viral cocktail' killing hares in UK and Ireland, scientist warns," *The Guardian*, August 27.

Baudrillard, Jean (1993) *The Transparency of Evil*, Verso.

Bauer, David L.V. (2021) "As a virologist I'm shocked my work has been hijacked by anti-vaxxers," *The Guardian*, September 7.

Belanger, Ashley (2022) "Facebook could be sued for addicting children under California bill," *ArsTechnica*, June 29.

Bergstroöm, Ulrika (2021) "New, active viruses found at depths of over 400 meters," *Phys.Org*, March 9.

Berkowitz, Reed (2020) "A game designer's analysis of QAnon: Playing with reality," *Curioserinstitute*, September 30.

Besteman, Catherine (2019) "Afterword: Remaking the world," Catherine Besteman and Hugh Gusterson, eds. *Life By Algorithms: How Roboprocesses Are Remaking Our World*, University of Chicago Press, pp. 165-180.

Besteman, Catherine and Hugh Gusterson, eds. (2019) *Life By Algorithms: How Roboprocesses Are Remaking Our World*, University of Chicago Press.

Bin, Chris and Joel Schectman (2024) "Pentagon ran secret anti-vax campaign to undermine China during pandemic," *Reuters*, June 14.

Birchall, Clare and Peter Knight (2023) *Conspiracy Theories in the Time of COVID-19*, Routledge.

Black, Finn (2022) "Against the urgency of normal," *Synapse: UCSF Student Voices*, February 7.

Boccaccio, Giovanni (2013) *The Decameron*, trans. Wayne Redhorn, Norton.

Boler, Megan and Elizabeth Davis, eds. (2020) *Affective Politics of Digital Media: Propaganda By Other Means*, MIT Press.

Boler, Megan and Elizabeth Davis (2018) "The affective politics of the 'post-truth' era: Feeling rules and networked subjectivity," *Emotion, Space and Society* 27, pp. 75-85.

Borreguero, Eva (2020) "El coste de la negación," *El País*. March 25.

Bouska, Josef (2020) "How Europe and America blew it on the pandemic: A tale of blindness and arrogance," *Salon*, October 22.

Bratton, Benjamin (2021) *The Revenge of the Real: Politics for a Post-pandemic World*, Verso.

Brin, David (1990) *Earth*, Bantam Spectra.

—— (1998) *The Transparent Society: Will Technology Make Us Choose Between Privacy and Freedom?* Perseus Books.

Burroughs, William (1967) *The Ticket That Exploded (The Nova Trilogy # 2)*, Grove Press.

Cadigan, Pat (1987) "Pretty boy crossover," *Patterns*, Tor, pp. 129-138.

—— (1991) *Synners*, Bantam Books.

Campbell, Denis (2021) "Scientists abused and threatened for discussing Covid, global survey finds," *The Guardian*, October 13.

Camus, Albert (1975) *The Plague*, trans. Stuart Gilbert, Vintage.

Centola, Damon (2018) *How Behavior Spreads: The Science of Complex Contagions*, Princeton University Press.

Cepelewicz, Jordana (2021a) "The hard lessons of modeling the coronavirus pandemic," *Quanta Magazine*, January 28.

—— (2021b) "Chasing the elusive numbers that define epidemics," *Quanta Magazine*, March 22.

—— (2021c) "DNA has four bases: Some viruses swap in a fifth," *Quanta Magazine*, July 12.

Chappell, Bill (2022) "What we know about the deadliest U.S. bird flu outbreak in history," *NPR All Things Considered*, December 2.

Chen, Caroline (2021a) "Why opening restaurants is exactly what the corona virus wants us to do," *Salon*, February 19.

—— (2021b) "Why the U.S. hasn't stopped syphilis from killing babies," *Undark*, November 23.

Christakis, Nicholas A. (2020) *Apollo's Arrow: The Profound and Enduring Impact of Coronavirus*, Little, Brown Spark.

Clynes, Manfred and Nathan Kline (1995) "Cyborgs in space," Chris Hables Gray, Steven Mentor, and Heidi Figueroa-Sarriera, eds. *The Cyborg Handbook*, Routledge, pp. 29–34.

Cockbill, Louisa (2020) "Public should be informed of uncertainties in model predictions of COVID-19 spread, say researchers," *physicsworld*, May 18.

Collins, Chuck and Omar Ocampo (2021) "Global billionaire wealth surges $4 trillion over pandemic," *Institute for Policy Studies*, March 31.

Criad, Miguel Ángel (2022) "Climate change is increasing the risk of interspecies viral contagions," *El Pais*, May 3.

Crichton, Danny (2021) "Everything is accelerating in the exponential age," *Techcrunch*, October 10.

CrimethInc. Collective (2020) "Surviving the virus: An anarchist guide," *CrimethInc.*, March 18.

Cubitt, Sean (2011) "Vector politics and the aesthetics of disappearance," John Armitage, ed. *Virilio Now*, Polity, pp. 68–91.

Cullen, Ben (1998) "Parasite ecology and the evolution of religion," Francis Heylighen, ed. *The Evolution of Complexity*, Kluwer Academic.

D'Angelo, Chris (2022) "Climate change is supercharging most infectious diseases, new study shows," *HuffPost*, August 8.

Davidson, Helen (2023) "Eight in 10 people in China caught Covid since early December, say officials," *The Guardian*, January 23.

Davis-Floyd, Robbie and Joseph Dumit, eds. (1998) *Cyborg Babies*, Routledge.

Davis, Hanna E. et al. (2023) "Long COVID: Major findings, mechanisms and recommendations," *Nature Reviews Microbiology*, January 13.

Davis, Mike (2020) *The Monster Enters: COVID-19, Avian Flu and the Plagues of Capitalism*, OR Books.

Dayen, David (2021) "The problem with Facebook is fraud," *The American Prospect*, November 2.

DeForest, Anna (2020) "The new stability," *New England Journal of Medicine*, 383, pp. 1708-1709, October 29.

Demuth, Bathsheba (2020) "The empty space where normal once lived," *The Atlantic*, August 28.

Dery, Mark (1996) *Escape Velocity: Cyberculture at the End of the Century*, Grove Press.

de Rivera, Javier (2020) "A guide to understanding and combatting digital capitalism," *tripleC: Communication, Capitalism & Critique. Open Access Journal for a Global Sustainable Information Society*, 18/2, pp. 725-743.

Derysh, Igor (2020) "Russia trying to 'incite violence by white supremacists groups' in US ahead of 2020 election: Report," *Salon*, March 10.

Devega, Chauncey (2021a) "The GOP's Ayn Rand death cult: Trump's party is killing the American people," *Salon*, February 24.

—— (2021b) "Vaccine paranoia: Why right-wingers are worried about their 'precious bodily fluids,'" *Salon*, August 9.

—— (2023) "The power of a conspiracy theory—and a 3-step plan to deprogram American idiocracy," *Salon*, January 23.

Devereaux, Ryan (2020) "Leaked documents show police knew far-right extremists were the real threat at protests and not 'AntiFa,'" *The Intercept*, July 15.

Devin, Keith (2020) "Can we really understand exponential growth?" *MAA-Mathematical Association of America*, June 1.

Díaz, Maria (2017) *Juegos de realidad alternativa: un análisis geonarrativo y aumentado de Ingress*, Doctoral thesis, Universidad Complutense de Madrid).

Dicker, Ron (2024) "Ex-CDC director makes alarming bird flu prediction," *HuffPost*, June 17.

Duprex, W. Paul et al. (2015) "Gain-of-function experiments: Time for a real debate," *Nature Reviews Microbiology*, 13, pp. 58-64.

Ehrenreich, Barbara (1989) "Foreword," Klaus Theweleit, *Male Fantasies, Volume 1*, University of Minnesota Press, pp. ix-xvii.

Eggers, David (2013) *The Circle*, McSweeney's Books.

Einenkel, Walter (2022) "Qronicles of Anon: COVID-19 vaccine is snake oil but injecting old urine into your arm is...good?" *Daily Kos*, April 23.

Erill, Evan (2022) "Viruses may be 'watching' you," *Salon*, September 21.

Fasce, Angelo (2020) "Prefacio: The upsurge of irrationality: post-truth politics for a polarised world / El recrudecimiento de la irracionalidad: políticas de la posverdad para un mundo polarizado," *Disputatio: Philosophical Research Bulletin*, 9/13, pp. 0–5.

Farah, Troy (2022) "Pandemic greed killed more than 1 million people globally, study says," *Salon*, November 8.

Fauzia, Miriam (2021) "Fact check: COVID-19 vaccines won't cause a zombie apocalypse," *USA Today*, April 14.

Featherstone, Mark and John Armitage (2020) "Exploring 'viral culture' " guest post, Duke University Press.

Ferguson, Cat (2021) "Meet the people who warn the world about new Covid variants," *MIT Technology Review*, July 26.

Fernandez, Anne Lutz and Catherine Lutz (2019) "Roboeducation," Catherine Besteman and Hugh Gusterson, eds. *Life By Algorithms: How Roboprocesses Are Remaking Our World*, University of Chicago Press, pp. 44–58.

Fielitz, Maik and Holger Marcks (2020) *Digitaler Faschismus: die sozialen Medien als Motor des Rechtsextremismus*, Dudenverlag.

Ford, Paul (2022) "Everybody's working for the good-binge," *Wired*, May, pp. 12–13.

Foucault, Michel (1979) *Discipline & Punish: The Birth of the Prison*, Vintage Books.

Galloway, Alexander (2004) *Protocol*, MIT Press.

Gammon, Katharine (2020) "The deadly plague that could devastate the US rabbit population," *The Guardian*, July 15.

Geary, James (2007) *Geary's Guide to the World's Great Alphorists*, Bloomsbury.

Geddes, Linda (2021) "Mathematicians discover music really can be infectious: Like a virus," *The Guardian*, September 22.

Gehl, Robert W. (2019) "Emotional roboprocesses," Catherine Besteman and Hugh Gusterson, eds. *Life By Algorithms: How Roboprocesses Are Remaking Our World*, University of Chicago Press, pp. 107–121.

Gest, Justin (2022) "How demographic shifts fueled by covid delivered midterm wins for Democrats," *Politico*, December 8.

Gibbs, Nancy (2009) "The case for modesty, in an age of arrogance," *Time Magazine*, November 9.

Gibson, William (1987a) *Count Zero*, Ace.

—— (1987b) "Johnny Mnemonic," *Burning Chrome*, Ace, pp. 1–22.

—— (1987c) "New Rose Hotel," *Burning Chrome*, Ace, pp. 103–116.

——— (1987d) "Burning chrome," *Burning Chrome*, Ace, pp. 168–191.

——— (2003) "The road to Oceana," *The New York Times*, June 25.

Gilbert, David (2021) "Facebook is making millions off lies about the climate crisis," *Vice*, November 5.

Goh, Irving (2021) "Virus is other people," *Cultural Politics* 7/1, pp. 145–149.

Goodman, Brenda (2022a) "A highly changed coronavirus variant was found in deer after nearly a year in hiding, researchers suggest," *CNN Health*, March 3.

——— (2022b) "Study reveals how Covid-19 infections can set off massive inflammation in the body," CNN, April 6.

Gore, Adrian (2022) "We need to let go of the Bell curve," *Harvard Business Review*, January 14.

Gottlieb, Scott (2021) "A second major seasonal virus won't leave us any choice," *The Atlantic*, September 12.

Graeber, David (2011) *Debt: The First 5,000 Years*, Melville House.

——— (2021) "After the Pandemic, we can't go back to sleep," *Jacobin*, March 4.

Graeber, David and Maja Kantar (2020) "Debt, bullshit jobs, and political self-organization," Renata Ávil and Srecko Hovat, eds. *Everything Must Change*, OR Books, pp. 217–230.

Graeber, David and David Wengrow (2021) *The Dawn of Everything: A New History of Humanity*, Macmillan.

Grant, Nico (2022) "Google's ethical AI team is still struggling," *Bloomberg*, February 4.

Gray, Chris Hables (1985) "The vaccines are coming," unpublished ms., History Consciousness Board of Studies, University of California at Santa Cruz.

——— (1997) *Postmodern War: The New Politics of Conflict*, Guilford.

——— (2001) *Cyborg Citizen*, Routledge.

——— (2014a) "Mind control: Burning books, burning bodies, burning minds" for "Auto de Fé" art performance by Hans Nevídal, at the Deutschen Nationalbibliothek, Frankfurt, Germany.

——— (2014b) "Big Data, actionable information, scientific knowledge and the goal of control," *Teknokultura*, 11/3, pp. 529–554.

——— (2005) *Peace, War and Computers*, Routledge

——— (2019) "Essay review: The threat of surveillance capitalism," *Teknokultura* 16, pp. 265–276.

——— (2025) *AI, Sacred Violence, and War: The Case of Gaza*, Palgrave.

Gray, Chris Hables and Ángel Gordo (2014) "Social media in conflict: Comparing military and social-movement technocultures," *Cultural Politics*, 10/3, pp. 251–261.

Greenspan, Rachel E. (2021) "Tucker Carlson appears to reference a COVID-19 conspiracy theory by claiming Bill Gates has 'powers' over our bodies," *Yahoo News,* February 23.

Gregory, Andrew and Alex Hern (2023) "AI poses existential threat and risk to health of millions, experts warn," *The Guardian,* May 10.

Grimes, Sara and Andrew Feenberg (2012) "Rationalizing play: A critical theory of digital gaming," Andrew Feenberg and Norm Friesen, eds. *(Re)Inventing the Internet: Critical Case Studies,* Sense Publishers, pp. 21–42.

Guffey, Robert (2020) Four-part series: "What is QAnon?" (August 16); "The deep, twisted roots of QAnon" (August 23); "Making sense of QAnon" (August 30); "Decoding QAnon" (September 7), *Cryptoscatology.*

——— (2022) *Operation Mindfuck: QAnon and the Cult of Donald Trump,* OR Books.

Gusterson, Hugh (2019) "Introduction: Robohumans," Catherine Besteman and Hugh Gusterson, eds. *Life By Algorithms: How Roboprocesses Are Remaking Our World,* University of Chicago Press, pp. 1–27.

Guterl, Fred, Naveed Jamali and Tom O'Connor (2020) "The controversial experiments and Wuhan lab suspected of starting the coronavirus pandemic," *Newsweek,* April 27.

Haas, Leonard and Sebastian Gießler (2020) "In the realm of paper tigers: Exploring the failings of AI ethics guidelines," *Algorithm Watch,* April 28.

Hall, Don and Chris Williams (2014) *Big Hero 6,* Disney.

Haraway, Donna (1995) "Cyborgs and Symbionts," Chris Hables Gray, Heidi Figueroa-Sarriera, and Steven Mentor, eds. *The Cyborg Handbook,* Routledge, xi–xx.

——— (2021) "The best possible now," Chris Hables Gray, Heidi Figueroa-Sarriera, and Steven Mentor, eds. *Modified: Living as a Cyborg,* Routledge, pp. 282–310.

Hartman, Thom (2021) "The deadly Omicron news for Trump-humpers laid bare," *Daily Kos,* December 19.

Hartnett, Kevin (2020) "The tricky math of herd immunity for COVID-19," *Quanta Magazine,* June 30.

Harvey, Josephine (2022) "Stacey Abrams enrages republicans by citing science on 'fetal heartbeats,'" *HuffPost,* September 22.

Haven, Janet (2022) "How to read the White House's blueprint for an AI Bill of Rights," *Data & Society Points.*

Heaven, Will Douglas (2021) "AIs that read sentences are now catching coronavirus mutations," *MIT Technology Review,* January 14.

Hecker, J.F.C (1888) *The Black Death and the Dancing Mania,* trans. B.G. Babington, Cassell & Company.

Henley, Jon (2021) "Is Russia's Covid vaccine anything more than a political weapon?" *The Guardian,* April 30.

Hester, Rebecca J. (2020) "Bioveillance: A techno-security infrastructure to preempt the dangers of informised biology," *Science as Culture* 27/1, March, pp. 153–176.

Holpuch, Amanda (2021) "US could have averted 40% of Covid deaths, says panel examining Trump's policies," *The Guardian*, February 11.

Honigsbaum, Mark (2022) "How to prevent the next pandemic by Bill Gates review: A germ of an idea," *The Guardian*, May 29.

Hvistendahl, Mara (2018) "You are a number," *Wired*, January, pp. 48–59.

Hvistendahl, Mara and Alexey Kovalev (2022) "Hacked Russian files reveal propaganda agreement with China," *The Intercept*, December 30.

Hyde, Marina (2024) "Hail Zuckus Maximus! The master of the metaverse is finally sorry… for being sorry," *The Guardian*, September 27.

Ismail, Kayla (2018) "AI vs. algorithms: What's the difference?" *CMSWire*, October 26.

Jana, Rosalind (2022) "The people who 'danced themselves to death'," BBC, May 12.

Jett, Jennifer and Cheng Cheng (2023) "China had almost 2 million excess deaths after the end of 'zero-Covid,' U.S. study finds," NBC News, August 25.

Jones, Steven E. (2009) "Second Life, video games, and the social text," *Publications of the Modern Language Association* 124/1, pp. 264–272.

Jordan, Tim (1999) *Cyberpower: The Culture and Politics of Cyberspace and the Internet*, Routledge.

Kabas, Marisa (2021) "People are eating horse paste to fight COVID: These doctors are one reason why," *HuffPost*, August 31.

Kaplan, Sarah (2016) "Mice quarantine themselves when sick—a lesson for us all," *The Washington Post*, August 23.

Karia, Sejal (2021) "Jesusalema: The dance sweeping the world from Angolans to Israeli monks and Irish police," ITV News, February 3.

Karlis, Nicole (2021a) "Why some New Age influencers believe Trump is a 'lightworker'," *The Guardian*, March 5.

—— (2021b) "Experts say the now-waning delta surge maybe the last major COVID-19 wave," *Salon*, September 23.

Kates, Graham (2020) "Twitter says fake 'Antifa' account was run by white supremacists," CBS, June 2.

Kelly, Kevin (1994) *Out of Control: The New Biology of Machines, Social Systems, and the Economic World*, Basic Books.

—— (2023) "Engines of wow," *Wired*, pp. 34–47.

Kenen, Joanne (2021) "How Covid-19 could make Americans healthier," *Politico*, February 18.

Kerr, Dara (2014) "On Facebook, good and bad moods are infectious," *c/net*, March 12.

Kim, Whizy (2022) "The major blind spot in Bill Gates pandemic prevention plan," *Vox*, May 10.

Klein, Colin, Peter Clutton and Adam G. Dunn (2019) "Pathways to conspiracy: The social and linguistic precursors of involvement in Reddit's conspiracy theory forum," *PLoS ONE* 14/11, November 18.

Klein, Naomi (2008) *The Shock Doctrine: The Rise of Disaster Capitalism*, Picador.

—— (2020) "Screen new deal: Under cover of mass death, Andrew Cuomo calls in the billionaires to build a high-tech dystopia," *The Intercept*, May 8.

Klippenstein, Ken (2021) "White supremacists, conspiracy theorists are targeting cell towers police warn," *The Intercept*, March 17.

Kolchinsky, Artemy and David Wolpert (2018) "Semantic information, autonomous agency and non-equilibrium statistical physics," *Interface Focus*. 8:20180041.

Kucharski, Adam (2020) *The Rules of Contagion: Why Things Spread—And Why They Stop*, Basic Books.

Lee, Ella (2022) "Fact check: Sculpture is evidence of antisemitic 'blood libel,' not false QAnon theory," *USA Today*, February 3.

Lee, Kai-Fu (2018) *AI Superpowers: China, Silicon Valley, and the New World Order*, Houghton Mifflin Harcourt.

Le Guin, Ursula (1968) *The Wizard of Earthsea*, Parnassus.

Levin, Sam (2018) " 'They don't care': Facebook fact checking in disarray as journalists push to cut ties," *The Guardian*, December 13.

Lewandowsky, Stephen, Ullrich Ecker, and John Cook (2017) "Beyond misinformation: Understanding and coping with the 'post-truth' era," *Journal of Applied Research in Memory and Cognition* 6/4, pp. 353–369

Lewis, Michael (2021) *The Premonition: A Pandemic Story*, Norton.

Lieser, Ethen Kim (2021) "Joe Biden is right: Facebook is 'killing people' says infectious disease expert," *1945*, July 19.

Losse, Kate (2018) "I was Zuckerberg's speechwriter: 'Companies over countries' was his early motto," *Vox*, April 16.

Lyons, E. J. et al. (2014) "Behavior change techniques implemented in electronic lifestyle activity monitors: A systematic content analysis," *Journal of Medical Internet Research* 16/8.

Macdonald, Helen (2014) *H Is for Hawk*, Grove Press.

Mackenzie, Deboray (2020) *COVID-19: The Pandemic That Never Should Have Happened, and How to Stop the Next One*, The Bridge Street Press.

Madad, Syra (2022) "Opinion: How a virus seemingly returned from the dead," CNN, August 15.

Mahdawi, Arwa (2014) "If Facebook is an infectious disease, here's a guide to the symptoms," *The Guardian*, January 23.

Mallapaty, Smriti (2021) "China's COVID vaccines have been crucial—now immunity is waning," *Nature*, October 14.

Malliagros, Thiago (2021) "Mega Anon vs. QAnon: Anatomy of a failed conspiracy theory and of a success," *The Skeptic*, December 20.

Marcotte, Amanda (2024) "Donald Trump's embrace of RFK Jr. Exposes the campaign's QAnon strategy," *The Guardian*, August 28.

Masco, Joseph (2019) "Ubiquitous surveillance," Catherine Besteman and Hugh Gusterson, eds. *Life By Algorithms: How Roboprocesses Are Remaking Our World*, University of Chicago Press, pp. 125-144.

Ma K. C., et al. (2024) "Genomic surveillance for SARS-CoV-2 variants: Circulation of Omicron XBB and JN.1 lineages—United States, May 2023-September 2024," *Morbidity and Mortality Weekly Report* 73, October 24, pp. 938-945.

Mann, Steven (2012) "McVeillance," *Veillance Blog*, October 10.

Mann, Steve, Jason Nolan, and Barry Wellman (2002) "Sousveillance: Inventing and using wearable computing devices for data collection in surveillance environments," *Surveillance & Society* 1/3, pp. 331-355.

Margaritoff, Marco (2024) "Elon Musk says his child is 'dead' to him in disturbing anti-trans tirade," *HuffPost*, July 23.

Markary, Marty (2021) "We'll have herd immunity by April," *Wall Street Journal*, February 18.

Masson, Victoria (n.d.) "Why is Eyam significant? *Historic UK*.

McBride, Lucy (2022) "We are not done," *Covid-19 Update Newsletter*, March 14.

McClure, Kelly (2022) "Study: COVIDS-19 infects four areas of the male genital tract," *Salon*, March 6.

McKenna, David (2016) "Eyam plague: The village of the damned," BBC, November 5.

McKenna, Maryn (2022) "Where did Omicron come from? Maybe its first host was mice," *Wired*, October 7.

McLuhan, Marshall (1996) "McLuhan on McLuhanism," Paul Beneditti and Nancy Dehart, eds. *Forward Through the Rearview Mirror: Reflections on and by Marshall McLuhan*, MIT Press, p. 40.

McNeill, William (1989) "Control and catastrophe in human affairs," *Daedalus*, 118/1, Winter, pp. 1-15.

Metzl, Jonathan M. (2022) "Foreword," Steven W. Thrasher, *The Viral Underclass: The Human Toll When Inequality and Disease Collide*, Celedon, pp. xi–xiv.

Middlemass, Keesha M. (2019) "A felony conviction as a roboprocess," Catherine Besteman and Hugh Gusterson, eds. *Life By Algorithms: How Roboprocesses Are Remaking Our World*, University of Chicago Press, pp. 77–87.

Milbank, Dana (2022) "Opinion: How does Ron DeSantis sleep at night?" *Washington Post*, March 14

Monod, Jaques (1971) *Chance and Necessity*, Alfred A. Knopf.

Monbiot, George (2018) "The Earth is in a death spiral: It will take radical action to save us," *The Guardian*, November 14.

Morens, David M., Peter Daszak, and Jeffery Taubenberger (2020) "Escaping Pandora's Box: Another novel coronavirus," *The New England Journal of Medicine* 3821, pp. 1293–1295, April 2.

Morton, Timothy (2015) "Introducing the idea of 'hyperobjects,'" *High Country News*, January 19.

Murphy, Gram and Lars Scheink (2018) "Introduction: The visuality and virtuality of cyberpunk," Gram Murphy and Lars Schmeink, eds. *Cyberpunk and Visual Culture*, Routledge, pp. xx–xxvi.

Naafs, Saskia (2018) "'Living laboratories': The Dutch cities amassing data on oblivious residents," *The Guardian*, March 1.

Nagle, Rebecca (2020) "Native Americans being left out of US coronavirus data and labeled as 'other,'" *The Guardian*, April 24.

Nasir, Arshan and Gustavo Caetano-Anollés (2015) "A phylogenomic data-driven exploration of viral origins and evolution," *Science Advances* 1/8, September 25.

Nelson, Ted (1974) *Computer Lib/Dream Machines*, Hugo's Book Service.

Nguyen, C Thi (2020) *Games: Agency as Art*, Oxford University Press.

Nicholson, Jonathan (2024) "Trump reposts 'Where we go one' QAnon rallying cry on social media," *HuffPost*, July 1.

Oamek, Paige (2024) "Elon Musk kicks off election day by going full QAnon," *The New Republic*, November 5.

O'Brien, Matt (2023) "Artificial intelligence raises risk of extinction, experts say in new warning," *HuffPost*, May 30.

O'Sullivan, Donie (2020) "Exclusive: She's been falsely accused of starting the pandemic. Her life has been turned upside down," CNN, April 27.

O'Sullivan, Donie and Konstantin Torpin (2020) "QAnon fans spread fake claims about real fires in Oregon," CNN, September 11.

Ouellette, Jennifer (2021) "Study: Folklore structure reveals how conspiracy theories emerge, fall apart," *arstechnica*, January 3.

Owen, Tess (2024) "Facebook is auto-generating militia group pages as extremists continue to organize in plain sight," *Wired*, October 29.

Paul, Ian Alan (2020) "Ten premises for a pandemic," ianalanpaul.com.

Paul, Kari (2021) "A few rightwing 'super-spreaders' fueled bulk of election falsehoods, study says," *The Guardian*, March 5.

Pearl, Robert (2020) "How the 80/20 rules can save your life during the coronavirus reopening," *Forbes*, May 26.

Plant, Sadie and Nick Land (1994) "Cyberpostive," Matthew Fuller, ed. *Unnatural: Techno-Theory for a Contaminated Culture*, Underground.

Pliny the Younger (2008) *Two Letters by Pliny the Younger to the Historian Cornelius Tactitus Regarding the Death of Pliny the Elder etc.*, trans. Kenneth Martin, Iron Bear Press.

Plummer, Kate (2024) "Donald Trump's COVID cure may have killed thousands," *Newsweek*, January 5.

Price, Polly (2022) *Plagues in the Nation: How Epidemics Shaped America*, Beacon Press.

PRRI (2023) "American values survey," Public Religion Research Institute (PRRI), October 25.

Quiggin, John (2010) *Zombie Economics: How Dead Ideas Still Walk Among Us*, Princeton University Press.

Reilly, Kaitlin (2017) "Why *The Circle* should have stuck to its original ending," *Refinery29*, May 1.

Reps, Paul and Nyogen Senzaki (1957) *Zen Flesh, Zen Bones: A Collection of Zen and Pre-Zen Writings*, Tuttle Publishing.

Resnick, Brian (2021a) "Why epidemiologists are so worried about the new Covid-19 variants, in 2 charts," *Vox*, January 8.

——— (2021b) "Psychiatrists are uncovering connections between viruses and mental health: They're surprising," *Vox*, December 1.

Rivas, Manuel (1988) *Books Burn Badly*, trans. Jonathan Dunne, Vintage.

Roberts, Paul (2018) "Russian fake news ecosystem targets Syrian human rights workers," *The Security Ledger*, May 11.

Robinson, Nathan J. (2022) "Why this computer scientist says all cryptocurrency should 'die in a fire'," *Current Affairs*, May 13.

Rose, Janus (2021) "Zuckerberg's Meta endgame is monetizing human behavior," *Vice*, November 1.

Rosenberg, Paul (2021) "A terrifying new theory: Fake news and conspiracy theories as an evolutionary strategy," *Salon*, August 8.

Ross, Andrew (1991) "Hacking away at the counterculture," Constance Penley and Andrew Ross, eds., *Technoculture*, University of Minnesota Press, pp. 107–134.

Rouvroy, Antoinette, and Thomas Berns (2013) "Algorithmic governmentality and prospects of emancipation," *Réseaux* 177/1, pp. 163–196.

Rozsa, Matthew (2021) "From deer and dogs to rats and mink, COVID-19 has spread to the animal world," *Salon*, September 8.

Rucker, Rudy (1986) "What is cyberpunk?" *REM*, #3, February.

Sample, Ian (2022) "Modern herpes variants maybe be linked to Bronze Age kissing, study finds," *The Guardian*, July 27.

Sastre, Paz Sastre and Ángel Gordo (2019) "El activismo de datos frente al control algorítmico. Nuevos modelos de gobernanza, viejas asimetrías / Data activism versus algorithmic control. New governance models, old asymmetries," *IC Revista Científica de Información y Comunicación* 16, pp. 157–182/183–208.

Scary, Elaine (1987) *The Body in Pain: The Making and Unmaking of the World*, Oxford University Press.

Schüll, Natasha D. (2014) *Addiction by Design: Machine Gambling in Las Vegas*, Princeton University Press.

Seneca, Lucius Annaeus (1833) "Which Treats of Earthquakes," *Natural Questions, Book VI*, trans. John Clarke, Loeb Classical Library.

Shah, Sonia (2016) *Pandemic: Tracking Contagions, From Cholera to Ebola and Beyond*, Picador.

Shanks, Michael (2015) "The future of archaeological theory: Looking forward with Ben Cullen," *mshanks.com*, December 29.

Shen, Chen, Derrick VanGennep, Alexander Siegenfeld and Yaneer Bar-Yam (2021) "Unraveling the flaws of estimates of the infection fatality rate for COVID-19," *Journal of Travel Medicine*, January 4.

Shortell, David, Christina Carrega and Josh Campbell (2020) "Vigilante group activity on the rise, worrying law enforcement and watchdog groups," CNN, August 30.

Siddique, Haroon (2020) "British BAME Covid-19 death rate 'more than twice that of whites,'" *The Guardian*, April 30.

Simon, Maite Fernández (2021) "'A woman is a woman and a man is a man': Putin compares gender nonconformity to the coronavirus pandemic," *The Washington Post*, December 22.

Sitaramana, Ganesh (2019) "Facebook threatens the economy, health and democracy," *The Guardian*, Feb. 24.

Smith, Adam (2021) "Conspiracy theorists spread '5G Covid' mind-control chip diagram that is actually a guitar pedal," *The Independent*, January 5.

Snowden, Edward (2021) "Why do conspiracy theories flourish? Because the truth is too hard to handle," *The Guardian*, July 1.

Spinney, Laura (2017) *Pale Rider: The Spanish Flu of 1918 and How it Changed the World*, Public Affairs.

Steinmetz, Katy (2018) "Why Slang is More Revealing Than You Realize," *Time*, December 12.

Stelter, Brian (2021) "The Facebook Papers consortium is growing, and reporters are gaining access to more documents," CNN, October 26.

Stephenson, Neal (1992) *Snow Crash*, Bantam Books.

Sterling, Bruce (1986) "Preface," in Bruce Sterling, ed. *Mirrorshades: The Cyberpunk Anthology*, Ace, pp. ix-xi.

—— (1999) "Cyberpunk in the nineties," Street Tech, *The Computer Lab/Areth Branwyn*.

Stillman, Jessica (2021) "He studied the behavior of 150 million people and found there are only eight reasons things go viral," *Inc.*, February 21.

Stobbe, Mike (2022) "How will COVID end? Experts look to past epidemics for clues," *HuffPost*, March 10.

Stout, Noelie (2019) "Automated expulsion in the U.S. foreclosure epidemic," Catherine Besteman and Hugh Gusterson, eds. *Life By Algorithms: How Roboprocesses Are Remaking Our World*, University of Chicago Press, pp. 31-43

Strabo (1903) *The Geography of Strabo*, trans. H.C. Hamilton, George Bell & Sons.

Subramanian, Samantha (2020) "The deep conspiracy roots of Europe's strange wave of cell-tower fires," *Politico*, May 18.

Sumner, Mark (2021) "Viruses do not get weaker over time and the COVID-19 pandemic is a prime example," *Daily Kos*, March 20.

Taylor, Steven (2022) "Pandemic psychology: Nothing new under the sun," *Knowable Magazine*, March 8.

Tech Transparency Project (2022) "Facebook profits from white supremacist groups," *Tech Transparency Project*, August 10.

Terrio, Susan (2019) "Detention and Deportation of Minors in U.S. Immigration Custody," Catherine Besteman and Hugh Gusterson, eds. *Life By Algorithms: How Roboprocesses Are Remaking Our World*, University of Chicago Press, pp. 59-76.

Tharoor, Ishaan (2023) "Elon Musk raises the specter of 'white genocide,'" *Salon*, August 2.

Theweleit, Klaus (1989) *Male Fantasies*, University of Minnesota Press.

Thompson, Clive (2020) "QAnon is like a game: A most dangerous game," *Wired*, September 22.

Thoreau, Henry David (2016) *Walden*, Macmillan.

Thrasher, Steven W. (2022) *The Viral Underclass: The Human Toll When Inequality and Disease Collide*, Celedon.

Thucydides (1982) *The Peloponnesian War*, trans. Richard Crawley, Modern Library.

Tréguer, Félix (2021) "The virus of surveillance: How the COVID-19 pandemic is fuelling technologies of control," *Political Anthropological Research on International Social Sciences (PARISS)* 2/1, pp. 16–46.

Tufekci, Zeynep (2017) *Twitter and Tear Gas: The Power and Fragility of Networked Protest*, Yale University Press.

Unsworth, Barry (1992) *Sacred Hunger*, Hamish Hamilton.

Valley (sic Vallely) Paul E. and. Michael A. Aquino (1980) "From PSYOP to Mind War: The Psychology of Victory," Headquarters, 7th Psychological Operations Group, United States Army Reserve, Presidio of San Francisco, California.

Villarreal, Luis (2008) "Are viruses alive?" *Scientific American*, August 8.

Vincent, James (2021) "Google is poisoning its reputation with AI researchers," *The Verge*, April 12.

Virilio, Paul (1999) *Politics of the Very Worst*, trans. M. Cavaliere, Semiotext(e).

Ward, Colin 1973. *Anarchy in Action*, Freedom Press.

Wark, Mckenzie (2019) *Capital Is dead: Is This Something Worse?* Verso.

Waters, Roger and Yanis Varoufakis (2020) "Snake oil or socialism?" Renata Ávil and Srecko Hovat, eds., *Everything Must Change*, OR Books, pp. 245–258.

Watson, Oliver, Gregory Barnsley, Jaspreet Toor et al. (2022) "Global impact the first year of COVID-19 vaccination: A mathematical modeling study," *Lancet: Infectious Diseases*, June 23.

WHO (2020) "What is dual-use research of concern?" WHO, December 13.

Wittgenstein, Ludwig (1922; 1998) *Tractatus Logico-Philosophicus*, trans. C.K. Ogden, Dover Publications.

——— (1989) "The big typescript," *Revue Internationale de Philosophies* 43/169, pp. 175–203.

Wood, Daniel (2021) "Pro-Trump counties now have far higher COVID death rates: Misinformation is to blame," *NPR*, December 5.

Wong, Julia Carrie (2020) "Down the rabbit hole: How QAnon conspiracies thrive on Facebook," *The Guardian*, June 25.

Wong, Tessa (2022) "Henan: China Covid app restricts residents after banking protests," BBC, June 14.

Wu, Katherine J., Ed Yong, and Sarah Zhang (2021) "Omicron is our past pandemic mistakes on fast-forward," *The Atlantic*, December 23.

Yang, Maya (2022) "Man who received landmark pig heart transplant may have died of pig virus," *The Guardian*, May 6.

Yong, Ed (2022) "We created the 'Pandemicene'," *The Atlantic*, April 28.

Youngblood, Mason (2020) "Extremist ideology as a complex contagion: the spread of far-right radicalization in the United States between 2005 and 2017," *Nature: Humanities and Social Sciences Communications* 7/49, July 31.

Zadronzny, Brandy (2021) " 'Carol's journey': What Facebook knew about how it radicalized users," NBC, October 22.

Zhang, Sarah (2021) "What if we never reach herd immunity?" *The Atlantic*, February 9.

Zimmer, Carl (2021) *A Planet of Viruses*, Third Edition, University of Chicago Press.

Zizek, Slavoj (2020) *Pandemic! COVID-19 Shakes the World*, OR Books.

Zuboff, Shoshana (2019) *The Age of Surveillance Capitalism: The Fight for a Human Future at the New Frontier of Power*, Public Affairs.

The Author

Chris Hables Gray lives in Santa Cruz, California. He is thes author of *Postmodern War, Cyborg Citizen, Peace, War and Computers,* and *AI, Sacred Violence and War—The Case of Gaza.* He is also the co-editor (with his dear colleagues Steven Mentor and Heidi Figueroa-Sarriera) of *The Cyborg Handbook* and *Modified: Living as a Cyborg.* He has also written or co-written several hundred peer-reviewed articles and book chapters in a dozen academic disciplines and fields, and just as many works of journalism and fiction.

An activist, gardener, and teacher, he aspires to finish his books-in-progress on violence, the serial killer Edward Wayne Edwards, evolution, information theory and California identity. And to get a cat.

Index

Abnormal(s), 16, 28, 173-175, 181, 188. *See also* Normal
Abouzeid, G., 133, 134
Accelerationism, 104-105
Acceptance, 75, 167, 179-180
Adrenochrome, 118
Affordances, 15, 43, 139, 145, 150, 166-171
Agency, 10, 27, 110, 150, 163
Age of Surveillance Capitalism, The (Zuboff), 5, 106, 108, 109, 137, 138-140, 142, 143-145, 148-149, 150, 161, 167-168
AIDS, 46-47, 181, 184, 191. *See also* HIV
Akufo-Addo, Nana Addo Dankwa, President of Ghana, 29-30
Albery, G., 15
Algorithm
 impact statements, 111
 iterations, 109
Algorithm Watch, 111
Algorithmic
 governance, 8, 110, 129
 governmentality, 111
 infection or contagion, 10, 36, 44, 46, 75, 138, 181
 society, 107-108
Algorithmic Intelligence (AI), 1, 18, 21, 20, 25, 36, 107, 108- 112, 135, 162, 185
Alternate Reality Games (ARGs), 122
Anarchism and anarchists, 126-127, 152, 165, 167
Anarchy in Action (Ward), 167
Anderson, L., 33
Anonymous, 149
Antigenic drift and shift, 34
Apophenia, 122-123
Arab Spring, 121, 149, 171
Arendt, H., 56
Aronson, E. and C. Tavris, 23, 83, 181
Arthur, W.B., 86

Artificial Intelligence (AI), *see* Algorithmic Intelligence (AI)
Artificial selection, 10, 174-175
Attack rate, 43
Attenborough, D., 96
Authoritarianism, 59, 68, 82, 90, 132, 170
Avian flu, *see* Bird flu
Azarian, B., 59
Azhar, A., 24

Bacigalupi, P., 153
Bannon, S., 120
Bats, 19, 23, 35, 50, 55, 60-61, 72
Baudrillard, J., 32, 77
Bauer, D.L.V., 81. *See also* Fake news
Behavior(s)
 collective/shared, 8, 95-97, 100, 168
 complexity, 25
 economic, 101
 manipulate, 27, 106, 113, 114, 129, 139, 143-144, 148-149, 157, 161
 online, 129, 160
 sickness, 94-95
 spread of, 21-22, 54-56, 59, 81, 89, 95, 100-102
 surplus, 144
Bell Curve, 29
Bell, D., 176
Benassi, M., 83
Berkowitz, R., 117, 122, 123
Berwick, Dr. D., 64
Besteman, C., 36, 107-108
 and Gusterson, H., 8
Bhairava, S. and M.V. Tantras, 178
Biden, J., 114, 177
Biden administration, 76-77, 111, 117
Big Other, 139-140, 144-145. *See also* Surveillance capitalism
Bing Nursery, 156-157

Index

Biological
  bodies, 8, 154
  contagion(s), epidemics, plagues and outbreaks, 22, 35, 43, 53, 56, 75, 78, 89, 90, 96, 104, 116–117, 183, 185
  evolution and change, 30, 57, 86, 87, 108, 174–175
  hunger, 136
  hyperobject, 13
  science of, 5
  war, 176, 190
Biological systems, 26, 86, 137, 153. See also Cyborg(s)
Biological viruses, 8, 19–20, 22, 31, 33–35, 99. See also SARS-CoV-2; Viruses of all types
  hosts, 40, 49
  and racism, 59–60
  as vectors, 56–57, 109
Birchall, C. and P. Knight, 88–91, 124
Bird flu, 57, 176
Bird watching, 32, 174, 193
Black Death, 23, 93, 95–96, 102, 151. See also Bubonic plague
Black, F., 177–178
Black Lives Matter (BLM), 5, 59, 65, 115, 125, 128, 171
Black swans, 29
Boccaccio, G., 8–9
Bodies, 5, 14, 26, 33, 35, 37, 87, 98, 99, 130, 186. See also Cyborg(s)
  bats and humans, 19, 55
  biological and political, 8
  burning, 63, 68
  dead, 93, 181
  hosts, 55
Body in Pain, The (Scary), 106
Bouazizi, M., 149
Bouska, J., 80
Bratton, B., 125, 127, 166
Brin, D., 168–169
Bubble, 6, 40

Bubonic plague, 84
Burroughs, W., 33

Cadigan, P., 160, 163–164, 170
Camus, A., see *Plague, The* (Camus)
Capitalism, 55, 72, 108, 126, 133, 135, 136–137, 145, 169, 180, 190. See also Data Disinfo Capitalism; Disaster capitalism; Surveillance capitalism; Vectorist class
  gangster and neoliberal, 82
  neo-capitalism in China, 57
  viral, 99, 140, 141, 150
Carson, C., 15
Case Fatality Rate (CFR), 43, 45
Case Rate (CR), 43
Catalysts, 6, 27, 42, 54–55, 57–60, 65, 190
Center for Disease Control (CDC), 44, 79, 93, 176, 191
Centola, D., 4, 46, 55–56, 70
Chance, 22, 28, 64, 66, 89, 98, 101, 134, 169, 179
  and necessity, 10, 86
Charters, E., 192
Chen, C., 9, 191
China, 2, 3, 41, 44, 50, 53, 57, 76, 78, 83, 88, 90, 111, 134, 140, 169–170, 182, 192
Cholera, 54, 70–71, 95, 165
Christakis, N.A., 23, 44, 79
Circle, The (Eggers), 141–143, 168–169
Civilization, 12, 15, 24, 26, 72, 80, 85, 98, 129, 142, 148, 174, 178
Climate
  change, 13, 15, 24, 57–58, 69, 72, 93, 148, 175, 179, 180
  collapse or destruction, 5, 8, 137, 151, 160, 163
  fiction, 152
  fostering disease(s), 72
  lies, 113, 115, 125
Clynes, M., 26, 154

Cognitive dissonance, 23, 65, 83, 181
Colonialism and pandemics, 30, 133, 134, 144
Community spread, 43
Complex contagions, 43, 56
Computer Lib/Dream Machines (Nelson), 167
Computer Professionals for Social Responsibility, 167
Computer viruses, *see* Digital viruses
Connor, J., 182
Conspiracies, 47, 63, 65–70, 72, 76, 79–81, 83, 84, 88–91, 106, 116–117, 130
Conspiracism, 66–67
Conspiracy
  entrepreneurs, 124
  practices, 66
  theories, 3–4, 65–66, 67–68, 84, 106, 118–119
Contact tracing, 44, 88
Contagion, 44, 56
Controls, 6, 54, 60–61, 65, 190
Cook, J., 120
Correlation
  does not prove causality, 44
  trading, 147–148
Corruption, 8, 58, 105, 112, 136, 140–141
Count Zero (Gibson), 159–160, 162
Courage, 16, 43, 60, 146, 174, 192
COVID-19, 1, 2–3, 9, 15, 80–81, 83, 88–91, 99, 132, 173–174, 176–177, 184, 191. *See also* SARS-CoV-2
  anarchist response, 126–127
  China's response, 44, 78, 83, 169
  conspiracies, 23, 26, 36–37, 55–56, 68, 69, 72, 80–81, 83, 88–91, 124
  costs, 59, 133, 150, 169, 190
  endemic, 1, 52, 184
  evolution of, 45, 92
  fatality rates, 14, 29, 42, 79–80, 132, 176, 192
  herd immunity, 45, 183
  hosts, animal, 35, 54, 60
  inequality and, 133–135, 190
  lessons from, 178–185, 191
  lockdown(s), 6–7, 17–18, 30, 39, 93, 96–97, 99, 123
  long, 28, 30–31, 35, 50, 69, 179
  origins, 72, 182
  pandemic, 2–3, 9, 13–14, 18, 51
  predictions, 2–3, 29, 35, 45, 182–184
  surveillance, 169
  "Roadmap for Living with COVID", 189
  science of, 47, 70–73, 177, 190, 192
  stress disorder, 179
  U.S. response, 79–80
  vaccines for, *see* COVID-19 vaccine
  variants, *see* COVID-19 variants
COVID-19 vaccine, 3, 11, 47, 52, 73–77, 79, 81, 89, 109, 189
  faith in, 13, 189
  fake news about, 36–37, 65, 67, 81, 128
  gain-of-function and production of, 73
  politics of, 30, 76, 134, 190
  variant targeted, 53
  viruses as well, 16, 61, 75
COVID-19 variants, 13, 28, 29, 45, 46, 50–51, 53, 75–76, 109, 163, 192
  Alpha, 52
  Delta, 51–52, 81
  naming, 40, 49–54
  new, 18, 43, 75, 183, 192
  Omicron, 25, 28, 42, 48, 51, 52–53, 78, 107, 170, 184, 192
  past of, 50
Credibility, 55–56
CrimethInc Collective, 126–127
Crypto currency, 148
Cubitt, S., 55
Cullen, B.S., 87
Cults, 4, 43, 123, 154, 162
Cultural
  change, 37, 58, 87–88, 130
  evolution, 86, 175

Cultural (*Cont.*)
  viruses, 20–21, 22–23, 31, 40, 58–59, 68, 77–84, 88. *See also* Social media, viruses and viral
Cultural Virus Theory, 3, 87
Culture, 65, 84, 86, 87, 98, 101, 130, 140, 167, 174
  digital and internet, 4–5, 25, 26, 122
  nature-culture distinction, 10–11, 91, 98
Cuomo, Andrew, Governor of New York, 79, 135
Currier, J., 82
Cyber attacks, 4–5, 149. *See also* QAnon
Cybernetics, 19, 26–27, 153, 179
Cyberpositive (Plant and Land), 140
Cyberpunk, 150–156, 158
  affordances, 166–171
Cyberspace, 121, 155, 158–160, 163–164, 166, 170
Cyborg(s), 7, 26–27, 77, 87, 98, 154, 164
CZU Lightning Complex fires, 63–64

D'Angelo, C., 57–58
Data & Society, 111
Data Disinfo Capitalism, 124
Davis, E., 58
Davis, H., 179
Davis, M., 19, 28, 34, 57, 74, 130, 146, 150
Davis-Floyd, R., 8
Dawn of Everything, The (Graeber and Wengrow), 87
Death, 37, 84, 94, 132, 136, 176, 178, 181, 187. *See also* Black Death
  evolution needs, 31, 86, 187
  fear of, 59, 94, 98
  fiery (pyroptosis), 35
  penalty or sentence, 103, 187
  rate(s), 52, 59–60, 78–80, 94, 165, 182
  Stoic death trip, 187. *See also* Seneca, L.
  threats, 83, 192
DeForest, Dr. A., 177
Democracy, 4, 54, 60, 78, 107–108, 111–114, 121, 130, 136, 141, 150, 190–191

Democracy Collaborative, 175
Demuth, B., 69
Denial of Death argument, 59
Dery, M., 163, 164
Devega, C., 36–37, 66, 82
Devin, K., 24
Dietz, K., 48
Diffusion of Innovation (Rodgers), 78
Digital capitalism, *see* Surveillance capitalism
Digital culture, *see* Culture, digital
Digital viruses, 20, 22–23, 31, 35–36, 40, 56–57, 82–83, 154. *See also* Algorithmic Intelligence (AI); Facebook; QAnon; Social media, viruses and viral
Disaster capitalism, 133–135, 150. *See also* Shock Doctrine; Surveillance capitalism
Disease X, 184
Doctorow, C., 106
Domestic violence, 104, 106
Dual use research of concern (DURC), 73
Dujarric de la Rivière, R., 34
Dumit, J., 8

Earhart, A., 31
Earth (Brin), 168–169
Earthquakes, 131–132, 180, 186–188
Ebola, 22, 48, 140, 184, 189
Economy, 29–30, 36, 113, 117, 134, 147, 188–189, 190. *See also* Capitalism
  carbon, 64
  desire or libidinal, 136, 139, 148
  e- or tele, 109, 135
  military, 135
  political, 58, 99
  surveillance, 148
  world, 133
Edmond J. Safra Center for Ethics, 189
Ehrenreich, B., 165–166
Election(s)
  of Biden, 67, 108
  of Trump, 1, 4, 118, 120
  misinformation, 48–49, 108, 115–116

Emotions, contagious or viral, 6, 23, 25, 54, 55–56, 58, 100, 112, 116, 129, 140, 144
Empathy, 56, 58
Endemic, 1, 44–46, 52, 123, 177, 184
Epidemic(s), 32, 44, 46, 52, 64, 79, 92, 100, 192. *See also* COVID-19; Plague(s)
  conspiracy theories, 79, 84
  diseases and climate, 58
  ending, 48
  financial crisis like SARS, 146
  Globisity, 100
  HIV, 14, 52. *See also* AIDS
  psychic (choremania), 95–96
  songs as, 100–101
  violence as, 104
Essential workers, 30, 40, 132, 189. *See also* Viral, underclass
Evolution, 10, 22, 46, 85–88, 91–92, 175
  and death, 31, 86, 187
  biological, 30, 57, 86, 87, 108, 174–175
  COVID-19, 45, 92
  cultural, 86, 175
  participatory, 86, 175
  technological, 86
  viral, 85
  viruses, 22
Excess mortality, 44. *See also* Death, rate(s)
Exploits, 44
Exponential
  age, 24
  change, 15, 105, 152
  curves, 5
  fire, 64
  growth, 17, 20, 24–25, 109, 176
  life cycle of viruses, 25
  network(s), 45
Eyam Hypothesis, 93–94

Facebook, 36, 102, 105, 112–114, 116, 119, 124, 142, 149, 160. *See also* GAFAM; Social media
Fake news, 18–19, 58, 81, 101–102, 115–116

Fascism, 6, 31, 41, 65, 82, 112, 152, 164–166, 181, 183
Fear(s), 6, 13, 35, 41, 56, 61, 69, 146, 159, 186. *See also* Empathy; Hope; Zombies
  and adrenochrome, 118
  and cognition, 58, 59, 61, 68
  of death, 59, 185
  identity and, 8, 37, 83, 124, 125
  infectious, 23, 59, 60, 102, 117
  monetizing, 124, 148
  my, 11, 50, 126, 142, 176, 179
  of normal (FONO), 12
  of the other, 68, 69, 82, 116, 181
  scary, 35, 51, 108, 138, 153, 154
  viral and virus, 18, 25
Featherstone, M. and J. Armitage, 32
Federal-State Fusion centers, 115
Feminism, 54, 152, 165
Fire(s), 69, 82, 98, 115, 170, 178, 186
  books, bodies and minds burning, 63, 68–69, 82
  cells inflamed, 35, 65
  wild, 63–65, 84
Flattening the curve, 28, 44
Flu , *see* Influenza/flu
Flynn, General Michael, 120
Folklore, 87–88
FONO (fear of normal), 12–13
Ford, P., 42
Forest, W., 2–3, 15
Foucault, M., 157–159, 160

Gaba, C., 79
Gain-of-function research, 72–73
Galloway, A., 43, 47
GAFAM (Google, Apple, Facebook/Meta, Amazon, and Microsoft), 108, 138, 141, 155, 160
Gamification, 111–112, 114, 116
Gaming, 111–112, 130
Gates, B., 36–37, 80, 88, 113, 135, 180, 189–190, 191
Gaussian distribution, 29

General Purpose Intelligence (GPI), 108–109. *See also* Algorithmic Intelligence (AI)
Generation interval, 45
Genetic susceptibility, 18, 45
Gibbs, N., 43
Gibson, W., 158–163
GISAID (Global Initiative on Sharing All Influenza Data), 51
Gladwell, M., 4
Goh, I., 96, 98–99
Google, 53, 105, 108, 111, 129, 135, 138–139, 141, 143–144, 155, 160. *See also* GAFAM; Social media; Surveillance capitalism
Gordo, A., 109
  and Gray, 1, 121
Gore, A., 29
Gottlieb, S., 184
Graeber, D., 129, 173, 178–179
  and D. Wengrow, 87
Grammar, 38, 49, 152
  viral, 21–27
Gray, C.H., 1, 26, 49, 73, 105, 109, 162, 163, 193
  and Gordo, 121
Gregory, A. and A. Hern, 1
Grimes, S. and A. Feenberg, 130
Guffrey, R., 119–120. *See also* Gamification
Gusterson, H., 110, 129
  and Besteman, C., 8
Guterres, Antonio, UN Secretary General, 134

Haddix, D., 79
Haldane, A., 146
Haraway, D., 26–27, 181–182
Hecker, J.F.C., 95–96
Herd immunity, 45, 74, 75, 183–184
Heteroglossia, 91
History, 7, 9, 14, 18, 50, 88, 92, 102, 113, 117, 130, 152, 178
  accelerating, 6

Arab, 134
bird flu, 176
capitalism, 140
deep, 30
pandemics and plagues, 22, 92–96. *See also* AIDS; Cholera; Influenza/flu
punk, 153
respiratory viruses, 184
of swords, 85–86
of technology, 86
HIV, 14, 46–47, 52, 76. *See also* AIDS
Homosexuality virus, 82
Hope, 6, 16, 33, 42, 69, 133, 150, 152, 155, 156, 165, 176, 177, 181, 183, 192
Hosts, 15, 19, 20, 22, 27, 40, 45, 54–55, 60, 93, 141, 190
  cultural, 37
  growth mediums, 40, 65, 109
How to Prevent the Next Pandemic (Gates), 189
Hyperobjects, 13, 15

Icke, D., 124
Idle No More, 5, 121, 149, 171
Index case, 45
Inequality, 13–14, 29, 55, 133–135, 189–190. *See also* Viral, underclass
Infection(s), 5, 45, 104, 148, 184
  avian/bird flu, 57
  digital, 105, 138
  controls, 47
  crypto currency, 148
  cognitive, 68
  cultural viruses, 22
  digital, 75
  fascism, 82
  foreign, 83
  information is, 22
  long COVID, 31, 52
  MRSA (*Staphylococcusaureus*), 71–72
  system effect, 23
  viral, 124

Infection Fatality Rate (IFR), 45, 52
Infection rates, 29, 32, 44–45,
    100, 106
  case, 43
  dependent happening, 48
  for herd immunity, 183
  reproduction (R), 47–48
Influenza/flu, 18, 45, 48, 50, 74, 173, 178,
    184, 192
  avian/bird, 57, 150, 176
  China, 83
  GISAID, 51
  long, 30–31, 192
  pandemic 1918, 18, 31, 34, 35, 42,
    57, 85, 92
  porcine, 57
  vaccine(s), 77, 192
  virus(es), 57
  vocabulary, 18
Infodemic(s), 41, 88
Information, 90, 143, 156, 167–168. *See also*
    Surveillance capitalism
  affordances, 166
  apophenia, 123
  control of, 8, 55
  digital and electronic, 55,
    158, 161
  economy, 158, 160, 168
  encoding, 103
  fake or nonsense, 66, 112, 116. *See also*
    Fake news
  infections through, 6, 22, 26, 45, 49
  meaning, 163
  personal, 78, 141, 158
  power, 106
  PSYOPs, 4
  systems, 9, 153, 167
  theory, 27, 40
  war, 4, 105, 121
Instagram, 36, 48, 124
Interconnection, 26. *See also* Networks;
    Information, systems

International Committee on the Taxonomy
    of Viruses, 20
Interweb, 106, 108, 116, 119, 121, 122, 128,
    154, 158, 167. *See also* Facebook; Social
    media; Twitter
  of things, 168
Ismail, K., 109

Jerez, A., 118
Jones, S., 122
Jordan, T., 157–159
Justice, 5, 12, 29, 65, 128–129, 130, 145,
    147, 165. *See also* Black Lives Matter;
    Idle No More; Inequality; Viral,
    underclass

Kelly, K., 167, 185
Keynes, J.M., 137
Kim, W., 189–190
Klein, N., 131, 133, 135, 150
Kropotkin, P., 126
Krugman, P., 36
Kucharski, A., 5, 20–21, 46, 48, 78, 81–82,
    100–102, 104, 146–148

Lamarck, J.-B. de, 86
Language of Virus, 7, 18–19, 32–38, 181
  grammar, 21–27
  rhetorical variants, 49–54
  'Ronaverse slang, 40–42
  technical dictionary, 43–49
"Law of Conservation of Catastrophe", 7
Lee, E., 118–119
Lee, K.-F., 110
Legitimacy, 56
Le Guin, U.K., 185
Life by Algorithms (Besteman and
    Gusterson), 110. *See also* Besteman, C.;
    Gusterson, H.
Life expectancy decline, 190
Links, 5, 18, 43, 45–46, 56, 101, 102, 161. *See
    also* Networks; Vectors

Lockdown, 6–7, 48, 133, 165, 174. *See also* Quarantine; Shelter-in-place
   anti-lockdown campaigns, 124
   Argentina, 39
   baking during, 17–18
   cult recruitment is similar, 123
   Ghana, 29–30
   New York sirens, 14
   in *Plague, The*, 32–33
   rituals, 96–97
   Santa Cruz, 32
   slang terms, 40–41
Long COVID, 28, 30–31, 35, 50, 52, 69, 80, 89, 94–95, 132, 179, 192
Losse, K., 113, 188

Macdonald, H., 32, 178
Mackenzie, D., 17, 24, 46, 93
Mahdawi, A., 114
Make America Great Again (MAGA), 59, 107, 117, 123
Malaria, 48, 77, 81, 140, 184
Male Fantasies (Theweleit), 165–166
Mann Gulch fire, 64
Mann, S., 155–156
Markary, M., 183
Mathewson, K., 115–116
Matrix, The, 151
May, R., 148
McBride, Dr. L., 12
McClendon, B., 143
McLuhan, M., 158–159
McNeill, W., 7
Media, 2, 9, 55, 58, 67, 89, 91, 130
   mass, 33, 104, 136, 142, 152, 158
   platforms, 57
   social, *see* Social media
Meta, 108, 113–114, 138, 141, 158, 160. *See also* Facebook
Metzl, J., 14
Milbank, D., 79
Military AI, 109, 135
Miller, J., 184

MILNET(s), 158, 167
Models, 22, 25–26, 37, 45, 88, 124, 146, 147, 184
   large language, 109. *See also* Algorithmic Intelligence (AI)
Mompesson, W., 93
Monod, J., 10
Monbiot, G., 24
Monster Enters, The (Davis), 150
Mora, Dr. C., 57–58
Mortimer, P., 28
Morton, T., 13
Mosquito theorem, 48. *See also* Reproduction rate (R)
MRSA (*Staphylococcus aureus*), 71–72
Music, 22, 96–97, 98, 108, 138, 144
   punk, 153
   viral songs, 100–101
Musk, E., 2, 107, 113, 120. *See also* Twitter; X
   DOGE, 117
   QAnon, 2, 120
Mutants and mutations, 10, 21, 25, 31, 46, 49, 52. *See also* Evolution
Myth of White Immunity, 128

Natural Questions (Seneca), 186–187. *See also* Seneca, L.
Nature-culture distinction, 10–11
Necessity, 10, 86. *See also* Evolution
Nelson, T., 167
Netherlands, 170
Networks, 22–23, 36–37, 46–47, 54, 56–57, 65, 113–114, 126. *See also* Links; Vectors
   corporate, 43, 109
   digitalized, 26, 35, 167
   military, 158, 167
   personal, 26, 36, 55, 56, 100
   resistance, 126
   small world topologies, 45
   strong-tie, 56
Nguyen, C.T., 112

Normal, 69, 151, 159, 173-175, 177-178. *See also* Abnormal(s)
    fear of normal (FONO), 12-13
    information spread, 22
    never again, 16, 28, 173-175
    new, 175
    next, 189
    politics of, 80, 108, 140, 177-178

O'Brien, M., 1
Omicron, 25, 28, 42, 48, 51, 52-53, 78, 107, 170, 184, 192. *See also* COVID-19
"One Planet, One Net" (Computer Professionals for Social Responsibility), 167
Online teaching, 12, 146-147
Operation Tunisia, 149
Optogenetics, 162
O'Toole, Á., 51

Pale Rider (Spinney), 18
Pandemic(s), 1, 2, 15, 17, 25, 49, 72-73, 89, 95, 152, 164, 180, 181, 184, 190
    1918 influenza, 92
    Age of, 72
    Black Death, 23, 93, 95-96, 102, 151
    cholera, 95
    COVID-19, 2-3, 9, 13-14, 18, 51
    of grief, 94
    psychology of, 179
    Shock Doctrine, 135. *See also* Shock Doctrine
    Warcraft, 154
Pandemicine, 15
Pandemic (Shah), 71-72
Pangolin(s), 23, 51, 55
Pango Network Lineage Designation Committee, 51
Panopticon, 157-160
Pareto curves, 29
Pascale, B., 120
Patient Zero, 46-47
Paul, I.A., 11, 127

Pearl, Dr. R., 29-30
Percentage(s), 6, 23, 47, 83, 89
Personal Protective Equipment (PPE), 41, 47, 76, 189
Phages, 34, 47
Plague(s), 5, 22, 39, 77, 84, 92-96, 117, 154, 165. *See also* Black Death; Contagion; Epidemic(s); Pandemic(s)
    Corrupted Blood plague, 154
    dance, 102
    generate conspiracies, 84
    history of, 92-94
    literature of, 178. *See also Plague, The* (Camus)
    politics and political, 27, 104, 125, 146, 150, 154, 183
    rules of, 5
Plague, The (Camus), 6, 8, 17, 32-33, 63, 92-93, 95, 102, 103-104, 117, 164, 178, 183
Planet of Viruses, A (Zimmer), 19
Plant, S. and N. Land, 140
Plasticity, 180
Platform capitalism, *see* Surveillance capitalism
Pliny the Elder, 188
Pliny the Younger, 185, 187-188
Police violence, 128-129. *See also* Inequality; Justice; Viral, underclass
Polio, 26, 30, 35, 74, 76, 184
Politics, 27, 30, 40, 50, 74, 127, 130, 132, 153, 165, 175. *See also* Anarchism and anarchists; Fascism
    accelerations, 104-105
    blowback into U.S., 4
    corruption of, 58
    of funding disease treatments, 191. *See also* AIDS
    and science, 18, 30, 76, 79, 191
    of social media, 112-117
Pompeii, 15, 135, 185-188
Pompeiicene, 15
Pop culture, 35

Post-truths, 116. *See also* Fake news
Power, 12, 13, 14, 50, 105, 113, 126,
    157–158, 167, 170, 178–179, 182. *See also*
    Inequality; Justice; Viral, underclass
    algorithmic, 8, 111, 144, 185. *See also*
    Surveillance capitalism
    corrupts, 140, 141, 163
    curves, 29
    defined, 105
    democratic, 190
    elites and, a, 99, 110, 111, 113, 125, 129,
        139, 155, 169
    of imagination, 193
    instrumentalist, 106, 139, 157–159, 163, 168
    knowledge is, 137, 156
    psychological, 119, 130
    seizure of, 107, 117, 120, 123, 129, 161
    soft, 157–159
    technological, 130. *See also* Social media
    to name, 50, 106
    vaccines as, 76
    of alliance, 167
Prasad, Dr., 177–178
Price, N., 115
Price, P., 84, 116–117
Prisons, *see* Panopticon
Privacy, 168–169. *See also* Circle, *The*;
    Panopticon; Surveillance capitalism
    is theft, 141
Protests, 54, 55, 77–78, 107, 125–126, 127,
    157, 170. *See also* Black Lives Matter;
    Idle No More
Protocol(s), 43, 47, 166, 167
PSYOPs (PSYchological OPerations), 4, 53,
    76–77, 119–120. *See also* QAnon
Punk, 153–155. *See also* Cyberpunk
Putin, V., 82, 136

QAnon, 1, 2, 3, 4, 13, 16, 44, 53, 81, 115,
    118, 122
    and AI, 129, 185
    distributed system, 167
    fake news, 115–116
    fantasies (alien lizards etc,), 2, 118, 120
    gamification, 112, 117, 119–123, 130
    influencers (moderators), 116
    and Kennedys, 120
    logic, 44, 121
    and MAGA, 117–125
    monetization (the grift), 123–124
    and Musk, 120
    origins, 53, 118, 123
    PSYOP, 4, 53, 76–77, 117–120
    resistance to, 130–131, 182
    social contagion, 101–102, 109
    and Trump, 1, 3, 4–5, 117–118, 120, 123, 185
    on vaccines, 81, 129–130
    virus, 37, 109
Quarantine, 2, 30, 40, 41, 60, 169, 93, 96, 173.
    See also Eyam Hypothesis; Controls
    defined, 39
Quiggin, J., 36

Rabbits, 54
    RHDV2 virus, 175–176
Racism, 1, 13, 55, 59–60, 99, 103, 110, 117,
    125, 128, 177, 181. *See also* Inequality;
    Viral, underclass
Redfield, R., 176
Reich, W., 112
Rent strikes, 127
Reproduction rate (R), 45, 47–48, 57, 100
Revenge of the Real, The (Bratton), 127
RHDV2 virus, 175–176
Rhetorical variants, 49–54. *See also*
    COVID-19 variants
Rivas, M., 63, 68–69, 82
Roboprocesses, 1, 8, 109–100, 110. *See also*
    Besteman, C.; Gusterson, H.
    reiterative algorithms for, 22, 36
    as zombification, 36
Rodgers, E., 77–78
'Ronaverse slang, 40–42
Ross, A., 167–168

Ross, R., 48
Rothkopf, D., 88
Rouvroy, A. and T. Berns, 111
Roy, A., 175
Rucker, R., 166
Rules of Contagion, The (Kucharski), 5, 20–21. *See also* Kucharski, A.
Rushkoff, D., 58
Russia, 4–5, 76, 112, 115, 134, 140, 180, 189

Sacred Hunger (Unsworth), 24, 131, 136, 190: *See also* Surveillance capitalism
Santa Cruz, 2–3, 6, 7, 11, 14–16
  Ben Lomond killings, 103
  earthquake in 1989, 131–132
  egg shortage, 176
  fires, 63–64
  hard lockdown, 2–3, 32
  howling with dogs, 97
  protests, 125–126
  sick in, 69, 173
  solidarity in, 180–181
  Tonga-Hunga Ha-apai tsunami, 28
  University of Santa Cruz, 11, 64, 126, 141, 146–147, 173
SARS-CoV-2, 1, 9, 10, 35, 50–51, 65, 94, 180–181, 189. *See also* COVID-19
  hosts, 54–55
  lies about, 68, 83
  life cycles, 28–31
  Long COVID, 35
  name, 50
  origin, 72–73
  spread, 46, 154
  testing, 49
  variants, 28–29, 48, 51, 192
  vectors, 55
  zombie making, 37
Scary, E., 106
Scher, E., 51
Schmidt, E., 135
Science, 5, 21, 79, 89, 129, 189, 190
  anarchists and, 126–127. *See also* Kropotkin, P.
  ancient, 71–72, 186–187
  COVID-19, 70–73, 146, 183, 190
  denial, 79, 80, 116, 120–121
  explained, 18, 26, 70, 71, 90–91, 127–128
  fiction (SF), 2, 108, 152
Security culture, 126
Seneca, L., 186–188
Sesame program, 169. *See also* China
Shah, S., 54, 57, 70, 71–72, 76, 77–78, 95, 140, 181, 184, 185
Shanks, M., 87
Shelter-in-place, 48. *See also* Lockdown; Quarantine
Shock Doctrine, 134–135. *See also* Klein, N.
Sickness behaviors, 94
Simple contagions, 56. *See also* Contagion; Links; Vectors
Singapore, 44
Sirius virus, 138
Slang, 40–42
Slutkin, Dr. G., 104
Small world topologies, 45. *See also* Links; Networks
Smart cities and high-tech dystopias, 135, 170
Smart phones, 138
Smith, A., 135
Smith, Agent (from *The Matrix*), 151
Smith, W., 34
Snow Crash (Stephenson), 151, 153–154, 156, 162
Snowden, E., 66, 161
Social distancing, 29, 48, 94, 99, 116. *See also* Controls
Social media, 2–5, 8, 19, 21, 36, 48, 53, 56–57, 67, 101, 102, 105, 109, 111–112, 121, 138, 139, 149, 161, 171, 185. *See also* Facebook; GAFAM; QAnon; Surveillance capitalism; Twitter; X
  AI's role, 124

Social media (*Cont.*)
  Big Other, 139. *See also* Surveillance capitalism
  Circle, The, 141-142
  gamification, 112-114
  influencers, 76, 89, 120, 123-124
  politics of, 112-117
  weaponization, 105, 117, 129
  viruses and viral, 58, 98-99, 114, 117
Social vectors, 55
Solidarity, 56, 67, 96, 102, 126, 127, 130, 180-181. *See also* Courage; Eyam Hypothesis; Hope
Solomon, S., 59
Sonalker, A., 135
Songs, 100-101. *See also* Music
Sophocles, 77
Spartacus, 188
Spinney, L., 18, 31, 34, 35, 42, 54, 60, 85, 93
Squared virality, 58. *See also* Viral, underclass
Stanford University, 17, 77-78, 107, 147, 156, 173
  Bing Nursery, 156-157
Steinmetz, K., 39
Stephenson, N., 151, 156. *See also Snow Crash*
Sterling, B., 152-153, 160, 170
Stocastic, 23-24. *See also* Percentage(s)
Stone, R., 120
Strategic complementarity, 55-56. *See also* Social vectors
Sugihara, G., 148
Suicides, viral, 104
Super-spreaders, 48-49, 58, 81, 89, 104. *See also* Catalysts
Surveillance capitalism, 108, 135-136, 159, 161. *See also Circle, The*; Disaster capitalism; Veillance Society
  Age of Surveillance Capitalism, The (Zuboff), 106, 108, 137, 138-140, 142, 143-145, 148-149, 150

Swords, 85-86
Synners (Cadigan), 163-164

Tavakoli, J., 147
Taylor, S., 179-180
Technical dictionary, 43-49
Technological change, 15, 86, 175
Technology, 8, 58, 91, 108, 130, 140. *See also* Algorithmic Intelligence (AI)
  corporations, 108. *See also* GAFAM
  democratic control of, 129
  diffusion studies, 55
  digital, 5, 8, 12, 139, 149, 169. *See also* Roboprocesses
  evolution of, 86
  exponential growth, 5, 152, 176. *See also* Exponential
  information, 105, 135
  military, 105
  mind control, 162
  politics of, 168, 169, 170
  sin, 160
Tennessee Supreme Court, 21
Terminology, *see* Language of Virus
Terror Management Theory, 59. *See also* Death; Fear(s)
Testing, 43, 49, 52, 53, 126, 189
Thompson, C., 122
Thompson, H., 118
Thoreau, H.D., 173
Thucydides, 95
Time, viral, 24-25, 30, 42, 175, 179, 190
Tipping points, 12, 179, 192
Trasher, S.W., 6, 14, 55, 58, 128, 181. *See also* Viral, underclass
Trump, D., 49, 67, 80-81, 107, 108, 112, 115, 116
  defeat 2020, 80, 90-91, 108
  election 2016, 112
  second term, 111
  fake news, 40, 115, 117-119
  fear and, 59
  Musk and, 124

QAnon and, 1, 3, 4–5, 117–118, 120, 123, 185
Reelection 2024, 1, 182, 185, 189, 120
Russia's goals, 115, 120. *See also* Russia
science denial, 49, 76, 79, 80–81
social media and, 130
super-spreader of fake news, 49
Tuberculosis, 177, 184
Twitter, 2, 36, 48, 56, 102, 105, 115, 124, 169. *See also* Musk, E.; Social media; X
Twitter and Tear Gas (Tufekci), 149

Uncanny, 12, 32, 37
Unsworth, B., 131, 136
Urbanization, 57. *See also* Smart cities

Vaccines, 61, 73–75, 89, 136, 177, 184, 190. *See also* Controls
cholera, 95
COVID-19, *see* COVID-19 vaccine
cyborgization and, 73
Vallely, P., 119. *See also* PSYOPs; QAnon; Trump, D.
Variants, 31. *See also* Evolution; Vaccines
COVID-19, *see* COVID-19 variants
digital, 37, 81
Marxist, 127
Moore's Law, 5
rhetorical, 49
Varoufakis, Y., 141
Vectorist class, 55
Vectors, 5, 6, 27, 42, 54–56, 60, 65. *See also* Networks
digital, 75
inequality, 14. *See also* Inequality; Viral, underclass
super-spreader, 58. *See also* Super-spreaders
viruses as, 73
Veillance Society, 141, 155–156, 160–163. *See also* Circle, *The*; Mann, S.; Surveillance capitalism
Velocity, 179. *See also* Accelerationism

Vesuvius, 185–188
Violence, 102–106, 128–129
controlling, 60
decline of, 145
doctors threatened with, 192
police, 128
stochastic, 24
viral, 20, 101, 102–106, 184
Viral
anthropocentrism, 98–99
emotions, 6
grammar, 21–27
life cycles, 27–32
principles, 5–6
racism, 99
sex, 34
time, 24–25, 30, 42, 175, 179, 190
underclass, 14–15, 164–165
vocabulary, *see* Language of Virus
Virilio, P., 55, 182
Virions, 49
Viruses, 9–10, 179
biological, 19–20, 22, 31, 34–35. *See also* SARS-CoV-2
hosts, 40
and racism, 59–60
as vectors, 56
bird flu, 57, 176
and climate change, 15
cultural, 20–21, 22–23, 31, 40, 59, 68, 77–84
Cultural Virus Theory, 87
and cyberpunk, 154–155
digital, 20, 22–23, 31, 35–36, 40, 56–57, 82–83, 138, 154. *See also* Algorithmic Intelligence (AI); Facebook; Media viruses; QAnon
language of, *see* Language of Virus
life cycles, 27–32
media, 58. *See also* Media
RHDV2, 175–176
and Veillance Society, 160–166
variants, *see* COVID-19 variants; Variants
von Pettenkofer, M., 70–71

Waitzkin, H., 190
Ward, C., 167
Wark, M., 55
Watkins, J., 120
Watts, C., 104
Wealth distribution, 40, 82, 110, 120, 133–135, 155, 161, 168, 188, 190. *See also* Inequality; Justice
Weber, M., 82
Wellcome Trust, 191
Wengrow, D., 87
Wenliang, Dr. L., 41–42
WHO (World Health Organization), 2, 51–52, 73, 76, 140, 191
Wittgenstein, L., 11, 21, 152
"Woke mind virus", 1–2. *See also* Musk, E.
World of Warcraft plague, 154

X, 2, 56. *See also* Musk, E.; Twitter
Xenoviruses, 3, 15

Zhang, S., 184
Zimmer, C., 19, 26, 28
Zizek, S., 136–137
Zombie Economics (Quiggin), 36
Zombies, 36, 136, 137
  zombification, 35, 36–37
Zuboff, S., 106, 108–109, 137–140, 142, 143–145, 148–149, 150, 161, 169. *See also* Surveillance capitalism
Zuckerberg, M., 113, 188. *See also* Facebook; GAFAM